Thinking About

THE UNThinkAble
in The 1980s

HERMAN KAHN

SIMON AND SCHUSTER
NEW YORK

Copyright © 1984 by the Hudson Institute, Inc.
All rights reserved
including the right of reproduction
in whole or in part in any form
Published by Simon and Schuster
A Division of Simon & Schuster, Inc.
Simon & Schuster Building
Rockefeller Center
1230 Avenue of the Americas
New York, New York 10020
SIMON AND SCHUSTER and colophon are registered trademarks
of Simon & Schuster, Inc.
Designed by Eve Kirch
Manufactured in the United States of America

10 9 8 7 6 5 4 3 2 1

Library of Congress Cataloging in Publication Data
Kahn, Herman, 1922–1983
    Thinking about the unthinkable in the 1980s.
    Includes index.
    1. Atomic warfare. 2. Deterrence (Strategy)
I. Title.
U263.K33   1984   355'.0217   84-1432
    ISBN 0-671-60449-X

CONTENTS

PART FOUR. MAKING THE WORLD SAFER

EDITOR'S NOTE

Herman Kahn died unexpectedly on July 7, 1983. At that time he had finished most of the material for this book, a project he was working on with Carol Kahn (not related), his long-time colleague and editor. Ms. Kahn completed the final manuscript with the assistance of the defense staff of the Hudson Institute.

INTRODUCTION
by BRENT SCOWCROFT

It is altogether fitting and wholly characteristic of Herman Kahn that even after he has left us he is making yet another important contribution to the dialogue on the most critical issue of our age. Those who have known Herman and received the benefit of his brilliant mind, fertile imagination and indefatigable energy would have expected no less.

Herman burst onto the national scene when he told us we *had* to think seriously about nuclear war and its consequences. It was he, in a sense, who "popularized" the national nuclear debate, notwithstanding the many important works that preceded *On Thermonuclear War* and *Thinking About the Unthinkable*.

Herman was one of the remarkable group of defense analysts who came out of the Rand Corporation in the latter part of the '50s and who were among the earliest definitive thinkers on nuclear warfare—the source of creative theories as well as practical applications of nuclear concepts. Herman was, in a way, the philosopher of the Rand group; while he was as concerned and knowledgeable as any systems analyst with the technical aspects of the nuclear threat and potential, his scope was always much broader, encompassing a humanitarian and ethical perspective. Are there moral and/or immoral implications of nuclear weapons? Can a nuclear war ever be justified? If so, how does it fit into the overall defense strategy of the United States? How does our defense strategy fit into our overall national objectives?

In many respects this work is an important—and timely—revisiting of *Thinking About the Unthinkable*. Again he underscores the essentiality of devoting every political and military effort to the goal of avoiding nuclear conflict. He points out the significance of nuclear weapons even if they are not used: the way in which they influence international politics, classic military doctrines, arms control negotiations and perceptions of power. Much of his thinking has been devoted to formulating "long-term anti-nuclear policies" that would help ensure that the weapons would not be used, policies that diminished the reliance on nuclear weapons even as they were predicted on a militarily strong United States.

But Herman forces us again to go beyond deterrence, to consider what we should do if deterrence fails, if nuclear weapons are actually used. He describes a number of measures that could be useful if a war did break out, including civil defense preparations (e.g. preplanned evacuations and a system of shelters) and ballistic missile defense. He argues vigorously that the time afforded by deterrence must be used not only to formulate a long-term approach to diminish the danger of war, but also to strengthen our position in the event deterrence fails. Herman had left us by the time the latest hypothesis—that a nuclear conflict could create a world-wide "nuclear winter"—was published. His reactions to this theory would most certainly have been enlightening. Whatever else he may have thought about it, I am confident he would not have concluded that his major theses had been invalidated.

The courage to confront the most horrifying aspects of these difficult issues has not always served Herman well. Some critics have accused him of being the original Dr. Strangelove. But while he does take on the moral simplicity of some of the ideas of anti-nuclear groups, anyone who knew Herman knows that his compassion and humanity, his deep morality and his patriotism, were as much a part of his nature as his toughness in addressing issues and his extraordinary and original intellectual scope. He really defied categorization—for example, he espoused a doctrine of "no first use" of nuclear weapons years before the "Gang of Four" made it fashionable.

My most recent encounters with Herman were at a series of so-called Breakfast Group meetings held over the past year in Washington. As a response to the current concern over the nuclear threat, Herman thought it very important for the Government to try to develop a consensus in support of a responsible long-term nuclear policy that, to a certain extent, would also take into account the specific fears expressed by the anti-nuclear groups. He brought together high-level military and civilian strategists and policy planners, people who could present some of the ideas raised by the Breakfast Group to the Defense Department and the President. Herman's motivation for organizing these meetings was simple—to do something for his country.

This book is very timely, and every American concerned about these vital issues should join Herman as he rethinks them. It is not yet clear who will—indeed, who can—carry on the task that Herman had set for himself. What does seem certain, however, is that it will be carried on; will be because it must be. In the Pentagon, in the State Department, in the Oval Office and in our living rooms, we have all become aware of the need to think about the unthinkable in the 1980s. For the role Herman Kahn has played in making it so and in helping us to do it intelligently, we are forever in his debt.

Thinking About the Unthinkable in the 1980s

PREFACE

"I can believe the impossible," Father Brown notes, in one of G. K. Chesterton's wonderful priest-detective stories, "but not the improbable." Unlike Father Brown, we believe not only the impossible *and* the improbable, but also the implausible, the unlikely, and the unproven. We believe in them and we take them seriously, especially when they involve what is probably the central issue of our time—nuclear war.

When I wrote *Thinking About the Unthinkable* in 1962, *not* thinking about nuclear war was still an option. It no longer is. Clemenceau's aphorism, that war is too important to be left to the generals, has been extended; clergymen, physicians, schoolchildren, and movie stars are no less entitled to express their concern about the nuclear threat than military analysts. The enormous range in the level of the public discussion has led to a great deal of misinformation and misunderstanding: much of today's conventional wisdom preaches the inevitability of disaster—an "apocalyptic vision" that only radical disarmament, or a "freeze," can prevent.

Fortunately, the impossible, as well as the improbable, implausible, unlikely, and unproven speculations about nuclear war (our scenarios as well as those of others) are not subject to reality testing. There are no direct precedents for the situation as it exists in the 1980s, or for extrapolations into the 1990s and beyond. But there are many important political, social, and historical precedents and continuities that can be brought to bear on the problems posed by nuclear weapons. In fact, we will argue later that these are as important as the discontinuities. For example, the Russians, and the Soviets, are and have been prudent, cautious, and nonadventurous in pursuing foreign policy objectives outside the Communist bloc. They fought wars that were mainly "thrust" upon them, and even though they managed to emerge each time stronger than before, they suffered greatly. We believe Soviet behavior is consistent enough to make plausible assumptions that preclude high-risk strategies, even if they had a significant probability of succeeding.

On the other hand, the Soviets have been obsessed with maintaining control over Communist satellite nations and would do almost anything to prevent the erosion of their dominant political authority. This fear, bordering on paranoia, has been factored into many of our nuclear scenarios. Nonetheless, we cannot provide the reader with any hardcore proof that either the assumptions or the concepts set forth here are entirely accurate. There are too many unknowns and uncertainties. What we can usually provide is evidence at the level of a "Scotch verdict."

In Scotland, a jury can find a defendant "guilty," "not guilty," or "not proven." If the jury chooses "not proven," it means it believes the defendant is guilty as charged but guilt cannot be proven beyond a reasonable doubt. We believe many of our arguments are valid, plausible, and credible but cannot absolutely prove them beyond a reasonable doubt. We are sure we are rarely, if ever, absolutely *wrong* on any issue, but the kind of documentation and theory to further substantiate our case so that it is "beyond a reasonable doubt" is often simply not available.

The point is that categorical references to right and wrong are often irrelevant (or certainly inadequate) criteria for determining the merit of many policies or discussions of nuclear war. Perception can be as important as reality; historical experience, culture, and ideology as important as abstract analysis and technical "facts." Thinking about nuclear war means thinking about the impossible, the improbable, the implausible, and the unlikely. But it also means making distinctions between scenarios that are more or less unlikely, more or less important, and more or less worrisome. In this context we often use the negative construction, "not implausible," "not incredible," and so on. Technically this formulation is called a "periphrasis." It is best illustrated quantitatively: take the term "not improbable." "Improbable" could normally be a probability of less than .1; "probable" a probability greater than .5; "not probable" less than .5. "Not improbable" is therefore greater than .1 and includes a much larger range than "probable." Some potential event can be judged "not improbable" and yet not be "probable." Similarly, a threat can be "not incredible" without being "credible." The need for earthquake-proof construction in regions in which earthquakes are rare but not completely improbable is illustrative of this concept.

The use of periphrases in strategic argument is central. We are dealing with remote (but momentous) events that most people find difficult to comprehend. We try to make the point that even though these possibilities are remote, they are not so remote, improbable, or implausible that they should not be taken with deadly seriousness. Indeed, it would not be the first time that history has turned out to be more imaginative and perverse than even the most fertile minds would have thought possible. The detailed "outbreak scenario" of World War I would probably have been rejected as the plot for a third-rate comedy of errors as simply too outrageous.* But the bizarre series of events did occur and brought with it enormous suffering.

Many of the concepts we deal with in this book are worth pursuing only because they are very much on the minds of concerned citizens everywhere. Some of the ideas are not on anyone's mind, but probably should be. All of them are presented in an effort to achieve a balanced and reasonable perspective on the often misunderstood dangers of nuclear war. To act responsibly we must learn as much as we can about the genuine risks involved in order to to make certain that we are better prepared to reduce them as much as possible. Or, deal with them.

---

* See Herman Kahn, *On Thermonuclear War* (Princeton, N.J.: Princeton University Press, 1960), pp. 350–75.

PART ONE

Thinking About the Unthinkable

1

An Overview *

The debate over nuclear war and national security policy is often more confused and confusing than informative and productive. In order to make it as useful and accurate as possible, it is important to separate issues that are relevant to government policy from issues that may be valid from some perspectives but are not serious policy options. Misconceptions and illusions do not contribute to the formulation of substantive recommendations and programs.

Twelve Nonissues

The following twelve assertions, however common and sincerely held, in terms of policymaking are basically irrelevant, impractical, inaccurate, or foolish and should be eliminated from the debate at the outset.

1. *We must halt the nuclear "arms race" in order to achieve the redemption of mankind.* This concept has recently been popularized in a book by Jonathan Schell:

* This is an introductory chapter which briefly surveys many of the more important topics addressed in the rest of the book. The Appendix serves a similar purpose, recapping the general contents of the book in outline form.

... today the only way to achieve genuine national defense for any nation is for all nations to give up violence together . . . if we had begun with Gandhi's law of love we would have arrived at exactly the same arrangement.

E. M. Forster told us, "Only connect!" Let us connect. Auden told us, "We must love one another or die." Let us love one another. . . . Christ said, "I come not to judge the world but to save the world." Let us, also, not judge the world but save the world.*

This concept has also been suggested in A Pastoral Letter on War and Peace by the National Conference of Catholic Bishops.† Redemption may be an appropriate and correct concern for a church, but it has nothing to do with any policies that the government can—or should—carry out. If there is a "redemption of mankind," it will not occur as a result of a great debate on national security policy or defense. It is, then, the "nonissue" of least relevance to government policy on nuclear war.

2. *The control of nuclear weapons should be pursued through the creation of an effective world parliamentary government and/or total worldwide disarmament.* A world government of sorts already exists: the UN Security Council. It is definitive on almost any issue on which a majority, including the five great powers (the United States, Soviet Union, China, Great Britain, and France), can agree. History, however, has proven its basic ineffectiveness.

The effort of trying to establish a more effective world government would itself involve major problems. In fact, it is very doubtful if the creation of the Security Council, or a successor body, could be negotiated today. The small nations intentionally turned over their power to the great nations of the world in 1945; today they fight fiercely to retain it by voting in blocks, changing allegiances and alliances as it suits their purposes. And there could be no chance of agreement today on who would have the veto power.

But even assuming a new world government were negotiable, how

* The Fate of the Earth (New York: Alfred A. Knopf, Inc., 1982), pp. 224 and 230.
† "The Challenge of Peace: God's Promise and Our Response," reprinted in Origins, May 19, 1983.

would it be structured? If it followed the "one man-one vote" principle, it would be dominated largely by Asian nations; under "one country-one vote," it would be run largely by the small states; and with "one dollar-one vote" it would result in domination by the United States, Japan, the Soviet Union, and Western Europe. All three options are unacceptable, and a compromise seems almost impossible. Under the circumstances, the Security Council is the best we have, and while a consensus by that body potentially could still have enormous impact (e.g., theoretically it is able to overrule national laws and possibly even the Constitution of the United States), it is unrealistic to expect a major and "binding" UN resolution on nuclear weapons to have much real-world significance.

As for total disarmament, there are almost 50,000 nuclear weapons in the world today; even if they were banned, not all would be destroyed. And even if they were destroyed, there is still a large amount of weapons-grade uranium and plutonium available, plus the knowledge of how to turn these materials into nuclear devices. It would not be acceptable to have a disarmament "solution" that allowed those with hidden weapons or weapons-grade material to gain an extraordinary advantage over the rest of the world.

3. *Even if it cannot be total, the goal should be disarmament rather than arms control.* The objective of nuclear-weapons policy should not be solely to decrease the number of weapons in the world, but to make the world safer—which is not necessarily the same thing. World War I broke out largely because of an arms race, and World War II because of the *lack* of an arms race. Similarly, many scenarios for the outbreak of nuclear war which are now "implausible" would become "not implausible" or possibly even "plausible" if the existing U.S. and Soviet nuclear arsenals were reduced to very small, less intimidating, and probably more vulnerable forces.

Developments that contribute to a safer world need not be the result of negotiations to reduce the number of nuclear weapons. For example, between 1967 and 1983, the number of weapons in the U.S. nuclear stockpile actually declined by 30 percent. The aggregate megatonnage in the stockpile declined by 75 percent between 1960 and 1983. These changes are attributable more to narrow military considerations (e.g., improvements in delivery accuracies diminished the utility of multimegaton warheads in many missions) than to arms-control influences. Nonetheless, the drop in the number of weapons and total explosive yield did lessen some of the dangers associated with nuclear war. Appropriate arms control could increase the trend toward decreased megatonnage and even toward fewer weapons, but unwise disarmament could set it back.

4. *There should be a total nuclear freeze.* A total nuclear freeze is counterproductive—especially now, when technology is rapidly changing and the Soviets have some important strategic advantages. An effective and verifiable freeze would worsen the U.S. position by "institutionalizing" apparent Soviet military superiority, and no freeze would be *reliably* verifiable, despite many claims to the contrary. More important, a freeze would prevent agreement on better arms-control measures by eliminating the incentive for useful negotiations and for the development of beneficial technology.

In fact, the most important reason for rejecting a freeze is that much of the weapons technology ahead is, relatively speaking, beneficial. Many people seem to believe that any change in weapons technology has to be for the worse, but that is demonstrably untrue. For example, almost all military analysts agree that the change from bombers to missiles in the early and mid-1960s made the world safer, since at that time missiles were much less vulnerable and accident-prone than the bombers. In addition, a freeze would preclude many adjustments and refinements that could make existing systems and forces safer. Nuclear-arms reductions or trade-offs are unlikely to take place under a freeze, but if our objective is to make the world safer, we must have the option of increasing, decreasing, or changing forces to achieve this goal.

As a first step toward a bilateral nuclear freeze, some antinuclear activists urge a unilateral halt to the development, testing, and deployment of U.S. nuclear forces. They believe this token of good faith will induce reciprocal restraint by the Soviet Union. Such support for a unilateral U.S. freeze betrays a gross ignorance of the history of the U.S.-Soviet arms competition for the last decade and a half.

From 1967 to 1983, the numbers of U.S. intercontinental ballistic missiles (ICBMs) and submarine-launched ballistic missiles (SLBMs) remained relatively constant, declining somewhat in the late 1970s and early 1980s. The number of U.S. strategic bombers decreased by 60 percent. During the same period, the number of Soviet ICBMs and SLBMs drew even with and then surpassed the inventories of U.S. ICBMs and SLBMs, while the size of the Soviet long-range bomber force grew slightly. In addition, the Soviets were very active in modernizing their strategic ballistic missile force. Over the sixteen-year period, the United States introduced two new or modified types of ICBMs; the Soviets fielded thirteen. The United States fielded two new or modified SLBMs; the Soviets deployed ten.

When in 1980 NATO withdrew one thousand theater nuclear weapons from Europe, in part to show good faith toward the upcoming negotiations on intermediate-range nuclear forces, no one paid much attention—least of all the Soviets. They increased, rather than decreased, the number of nuclear warheads trained on Western Europe. In short, the Soviets find it less than morally and psychologically compelling to match any unilateral U.S. efforts to dampen the arms competition. A unilateral freeze would most likely be exploited by the Soviets to augment their margin of military advantage over the United States and its allies.

But assuming a freeze were negotiated, it would not prevent either side from working on nonnuclear systems (e.g., modernized air defense networks, improved capabilities for antisubmarine warfare) which would not be frozen. Both parties would also have to worry about a technological "breakthrough" or a "breakout" by the other side.* Each side could justify its own attempt to achieve a breakthrough and then a breakout by "discovering" that the other side was cheating. New nuclear and nonnuclear systems could both be made extraordinarily effective against an opponent whose current systems were frozen.

As a political maneuver, the call for a nuclear freeze has turned out to be an effective way of telling national leaders in the West that *something* must be done to allay public fears of nuclear war. But it has also been counterproductive by sending Soviet decision makers the message that they can gain more by manipulating Western public opinion than by making genuine attempts at arms control. As a national policy, however, there is simply no case for a freeze.

5. *Deterrence must be made 100 percent reliable.* This is a nonissue simply because there is no way to make any complex human (let alone technical) system 100 percent effective.

---

* In general, "breakout" involves the relatively sudden, explicit, and unilateral abrogation of an arms-control agreement through the fielding of new or additional weapons that had been constrained by the broken treaty. With the illegal deployment, the violator hopes to gain significant military advantages or political leverage.

6. *Deterrence must fail eventually, and probably will fail totally.*
There is no way we can ignore this possibility. But many doomsayers
argue that since it *might* fail, it eventually *has* to fail. This is technically
incorrect, but not entirely unreasonable (unless it assumes that the failure must be total). My guess is that nuclear weapons will be used sometime in the next hundred years, but that their use is much more likely to
be small and limited than widespread and unconstrained. Deterrence
would then have failed—but not totally. This is why the following points
are so important.

7. *Useful "damage limitation" in a nuclear war is infeasible;* or
8. *One can achieve totally reliable damage limitation.* If counterforce
attacks and strategic defenses* could reduce damage in a nuclear war
to, perhaps, 10 million to 40 million deaths instead of the 50 million to
100 million that might occur in their absence, and if various other measures could improve the effectiveness and rapidity of recovery from a
nuclear war, then they would be worth the effort. Some people believe
that the development of such capabilities would be reckless because
they would encourage U.S. complacency about nuclear war and U.S.
risk taking in crises, and/or excessive Soviet fears which would lead
them to preempt a U.S. strike. Others believe systems and operations
intended for damage limitation must be made foolproof to be worthwhile.

There is no such thing here as total reliability, but if appropriate
efforts could mitigate the effects of a nuclear war, then perhaps even an
unprecedented catastrophe would not be a total disaster. There are
many contexts in which even marginally effective damage-limitation
programs might be very effective.

The usual discussion about counterforce wars and strategic defenses
does not recognize this in-between position, i.e., that some reduction is
better than none and that efforts to limit damage will not produce a
material increase in the likelihood of nuclear war. I would argue that it
is immoral for a nation not to take at least the relatively simple and
inexpensive measures it can to achieve an "improved war outcome"in
the event that deterrence fails.

---

* "Counterforce" attacks are strikes directed against the enemy's military assets (e.g.,
land-based missiles, bombers, command-and-control facilities). "Strategic defenses" are
systems providing protection against nuclear attack. Ballistic missile defense and air defense systems are "active" defenses; civil defense measures are "passive" defenses.

9. *A nuclear war can be reliably limited;* or 10. *There is no possibility of a limited war.* No one can guarantee that either of these predictions would accurately describe an actual nuclear war. In some circumstances, some kinds of limited nuclear war are clearly possible. There are very large and very clear "firebreaks" between nuclear and conventional war. In nuclear war, the primary firebreak might be "no attacks on the homeland," the next might be "no attacks on cities," and so on. Both sides may choose to observe these firebreaks, though no one can absolutely guarantee they will. But one also cannot be sure they will be totally flouted.

11. *There can be no victory in nuclear war (i.e., "nobody wins a suicide pact"),* or 12. *Either the United States or the Soviet Union could rely on victory.* These issues have become a morality test: to say that "a nuclear war might be limited" or that "there might be a victor in a nuclear war" is to label oneself as a nuclear war hawk (one who seeks to start a nuclear war) or a defender of similarly monstrous positions. To me, it is outrageous to make a morality test out of a realistic and important observation—and both of the quoted remarks are realistic and important.

It is incorrect to say that victory in nuclear war is impossible. It is especially possible if either side, or both, have low levels of nuclear forces that are vulnerable to destruction through creative or clever enemy tactics. Unfortunately, it is even quite conceivable at current levels of nuclear armament—but neither nation is likely to choose to go to war just because it has developed some ingenious war plan that might work; the risk is too frightening, and the present governments on both sides are too prudent and cautious. However, that differs from saying that victory could not happen. Indeed, some reasonable facsimile of victory could be achieved even by a nation forced into war. The Soviets, for example, won the war started by Nazi Germany despite suffering 20 million deaths and losing a quarter of their capital stock.

It is unwise to judge realistic analyses and preparations as a lust for war. There are no two sides to the nuclear debate: no one is "for" war; everyone is against it—some categorically so and others only to the degree that it does not result in an even less desirable alternative.

These twelve "nonissues" are important because many of them are deeply held beliefs and reflect genuine concerns. But they offer no substantive guidance in dealing with nuclear dangers, cannot be translated into constructive programs, and often stand in the way of serious discussions of useful and necessary programs.

Twelve Almost Nonissues

One step removed from these "nonissues" are twelve propositions that are equally popular, widely held, and emotionally defended, but offer almost as little practical guidance on how to make the world safer from the threat of nuclear war. The difference is that while we are comfortable in our belief that the first twelve are truly irrelevant, we are not as sure about these (i.e., they involve a "Scotch verdict").

1. *Nuclear war would result in the destruction of the created order,* and/or 2. *Nuclear war would result in the destruction of all human life.* There are no respectable objective analyses or calculations to indicate that either of these is likely.* But the data and theory are so lacking one cannot be absolutely certain. From a scientific perspective there is some indication that a nuclear war could deplete the earth's ozone layer or, less likely, could bring on a new Ice Age—but there is no suggestion that either the created order or mankind would be destroyed in the process.† From a religious perspective these assertions are almost heretical, since only God can fashion or destroy the universe (but some believe that He may choose nuclear war as His means of doing so). As a practical matter, however, these concerns do not offer any useful policy guidelines.

* Since Herman Kahn's death, new studies have emerged expressing concern over a possible "nuclear winter." The theory suggests the potential for a substantial drop in the surface temperature over land areas of the northern hemisphere (due to atmospheric smoke) following a relatively large-scale nuclear attack, especially on cities. Presumably, if more than a thousand cities were involved, a temperature drop (20°C–40°C) could occur that could, at least temporarily, turn summer into winter and have a severe impact upon human survivors as well as on plant and animal life. It is also considered possible that a less drastic but still important temperature drop would spread to the southern hemisphere.
  If the nuclear-winter theory turns out to be correct, it would have implications for some of the scenarios discussed in this book but would not appear to be significant for many of the limited or controlled war scenarios discussed here. We cannot suggest how Mr. Kahn would have modified the relevant parts of his analysis to accommodate this new threat, except to say that he surely would have taken any new factors into account.
  † One National Academy of Sciences study of a hypothetical 10,000-megaton war explicitly concluded that both the biosphere and man *would* survive. See *Long-Term Worldwide Effects of Multiple Nuclear-Weapons Detonations* (Washington, D.C.: National Academy of Sciences, 1975).

3. *The threat of a nuclear war would mean "everybody Red, dead, or neutral."* In Europe, when the only alternatives were presented as "Red" or "dead," the obvious choice was "Red." However, sophisticated Europeans are now formulating the choice to include "everybody neutral." A further revision of the slogan—"everybody Red, dead, neutral, *or NATO*"—is much more reasonable. Indeed, the purpose of the Atlantic Alliance is to make the last option the most attractive one.

4. *Nuclear weapons are intrinsically immoral.* Nuclear war is such an emotional subject that many people see the weapons themselves as the common enemy of humanity. Nuclear weapons are intrinsically neither moral nor immoral, though they are more prone to immoral use than most weapons. But they can be used to accomplish moral objectives and can do this in ways that are morally acceptable. The most obvious and important way is to use them or their availability to deter others from using nuclear weapons. The second—of much lower, but still significant priority—is to use them to help limit the damage (human, social, political, economic, and military) that could occur if deterrence fails. Anything that reduces war-related destruction should not be considered altogether immoral.

On the other hand, the position that nuclear weapons are "just another" weapon and therefore as moral as any other is not accurate either. I would judge them as moral when used solely to balance, deter, or correct for the possession or use of nuclear weapons by others, and immoral when deliberately used against civilians, for positive gains, or to save money and effort on nonnuclear military alternatives. This rule precludes the first use of nuclear weapons to defend Western Europe (current NATO policy), simply to avoid the more complicated capabilities, plans, and costs required for improved conventional defenses. It likewise stamps as immoral the targeting of enemy cities simply to avoid the problems of counterforce weapons and attack planning (countercity targeting is endorsed by many supporters of the nuclear freeze).

It is unacceptable, in terms of national security, to make nonuse of nuclear weapons the highest national priority to which all other considerations must be subordinated. It is immoral from almost any point of view to refuse to defend yourself and others from very grave and terrible threats, even as there are limits to the means that can be used in such defense.

5. *Expenditures for strategic nuclear forces are bankrupting the United States and the Soviet Union.* U.S. strategic expenditures are now less than one-tenth of 1 percent of the gross national product (and have been less than 2 percent for almost two decades), while Soviet expenditures are probably about 2 percent of the Soviet GNP. The cost of U.S. strategic forces represents only about 10 percent of total U.S. military expenditures; roughly 15–20 percent of the Soviets' defense budget is allocated to strategic forces. So while reducing the cost of arms is a desirable goal, it is simply not an overriding priority in terms of bringing about a dramatic turnaround in U.S. fiscal solvency.

6. *Defense expenditures should be reallocated to the poor.* A healthy and fully functioning society must allocate its resources among a variety of competing interests, all of which are more or less valid but none of which should take precedence over national security. This is not an argument for papering over instances of wasteful or excessive spending in the Department of Defense, but simply a recognition that national security programs have a legitimate and fundamental claim on the nation's resources. Furthermore, even if defense expenditures were cut, the savings would be divided along the lines of current federal fiscal allocations. Only those who are ideologically opposed to military programs think of the defense budget as the first and best place to get resources for social welfare needs.

7. *"War-fighting" measures are simultaneously too ineffective and too effective,* and 8. *"Deterrence only" is the least undesirable policy; any "war-fighting" policy is fatally flawed.* As used by most strategists, "deterrence only" implies an all-or-nothing strategy: first, a very strong belief that deterrence can and must be made to work for the foreseeable future, and second, that if it fails we are all doomed—because the cities on both sides would be deliberately and automatically destroyed at the outset of a war. The policy of "mutual assured destruction" (MAD) is the clearest form of "deterrence only." *

It is often believed that any attempt to mitigate the damage a nuclear war would cause, or even to reinforce deterrence with "war-fighting" capabilities, is "destabilizing" or a waste of resources. For its supporters, a "deterrence-only" policy offers a simple, clear and, by itself, unaggressive "solution" to the threat of nuclear war. But the term "war

---

* See section on "Useful terms," p. 40.

fighting" does not mean one *wants* to fight a war; the position simply recognizes that deterrence can fail and says that it is prudent to have programs both to reinforce deterrence and to alleviate such failure should it occur. As I will argue later in more detail, these efforts are likely to be effective enough to be worthwhile (i.e., they could have a positive effect on the course and outcome of a war), but would not be likely to cause or contribute to an appreciable increase in the risk of war.

9. *No significant weakening of deterrence is acceptable.* In most situations, some small "destabilizing" or weakening of deterrence is not likely to be significant. In some instances, an intentional weakening might be an acceptable part of a trade-off for other gains. For example, a minor and relatively insignificant decrease in deterrence would be justified if it would bring about an enormous reduction in the damage done if deterrence failed. Deterrence itself is not a preeminent value; the primary values are safety and morality.

10. *If retention of nuclear weapons is unavoidable, then "simplistic stability" is preferable to "multistability."* If we recognize nuclear deterrence as a means toward attaining a safer overall security environment, then *simplistic stability* (stability only against a first strike) should not be the sole objective of strategic forces. In fact, it is not the only mission of U.S. forces: their purpose is multistability—i.e., to deter serious provocations against the United States (and its allies), as well as to prevent an actual first strike.

Multistable deterrence imposes much more stringent—and necessary—requirements on our strategic forces than simple deterrence. While the U.S. can no longer have the kind of extended deterrence that covers all areas where provocation is possible, it still needs a "not-completely-incredible" ability to punish an opponent for a variety of (extreme) provocations. (Multistable deterrence is described at greater length in chapter 5.)

11. *Nuclear war would be fought mainly to achieve positive gains.* Both the Soviet Union and the United States are essentially very prudent and cautious—the Soviets probably even more than the Americans —and both are unlikely to risk a nuclear war for positive gains (i.e., to fulfill or advance national or personal ambitions), even if they think they can do it successfully. Neither the United States nor the Soviet Union

believes in "war by calculation" (i.e., planned and carried out as scheduled), but rather by miscalculation (plans always go awry). Calculated
wars have not worked in either country's history, leading both to the
realization that even theoretically sound plans are almost certain to go
astray. Consequently, only desperation could persuade the leadership
of either country to initiate a nuclear war.

The belief that war has become virtually obsolete as an instrument
for advancing positive ambitions (as opposed to averting disaster) is not
completely the result of nuclear deterrence. It is also because, increasingly, the most effective and reliable way to achieve affluence, power,
influence, status, and prestige is through economic development, and
not through territorial or political gains. While war between two major
powers may still be an important way for one or both of them to avoid
what appear to be more severe alternatives (e.g., for the Soviet Union,
the loss of political control), primarily it has become largely dysfunc-
tional. And yet, maintaining a credible (or "not incredible") capability
to resort to nuclear war remains basic to an adequate defense and foreign policy.

12. *Normally, there is an automatic and increasingly dangerous
"arms race."* This concept is plausible and has sometimes been accurate—for example, the arms race preceding the outbreak of World War
I. Post-World War II history has witnessed some automatic increases in
armaments. New developments in weapon systems during the 1950s and
early 1960s created a situation that was most dangerous, and even conducive to accidental war. In retrospect, we were fortunate that none
occurred.

However, it appears that in both the United States and the Soviet
Union the radical development of military technology since World War
II has been a greater engine of weapons development than has any sense
of detailed measure/countermeasure competition. A careful study of the
development of Soviet and American strategic weapons suggests that in
almost all cases weapons were developed in conformity with strategic
and technological thinking that took remarkably little account of the
other side's real or likely programs and other strengths and weaknesses.
If each side must more or less commit itself to a weapon system as
much as ten to fifteen years before it enters service in any quantity, then
there is little chance that the concepts being developed in one country
will serve as a guide for the other.

The world became much less dangerous in the 1960s because of improvements in equipment (e.g., submarine-launched ballistic missiles —SLBMs), tactics (e.g., alert procedures), and thinking (e.g., planning for limited counterforce campaigns). But even with advances in U.S. and Soviet weaponry and defense planning, there was no "arms race." For example, from 1963 to 1980, the U.S. defense budget was more or less constant in constant dollars (except for the operational expenses of the Vietnam War), while the Soviet budget increased by 4 or 5 percent a year—about the rate of increase for the Soviet GNP.

More accurate than the "race" metaphor is the observation that if it was a contest at all, the Americans walked while the Soviets trotted. There was no race—but to the extent that there was an *arms competition*, it was almost entirely on the Soviet side, first to catch up and then to surpass the Americans. The United States barely competed: except for some retrofitting (e.g., equipping ICBMs and SLBMs with multiple warheads), the U.S. defense establishment languished. The present controversy over the expense and morality of "rearming" the United States is a result of two decades of a very lax U.S. defense effort; the controversy could largely have been avoided if a consistent pace and pattern of defense preparedness had been maintained all along. The United States and the world would be much safer today, and the current anxiety-provoking defense program would not be necessary. But even this recent attempt to redress the balance is very different from being involved in an inevitable and increasingly dangerous "arms race." In fact, as discussed later, many of the innovations in defense that we expect over the next decade or two are as likely as not to make the world safer.

The accuracy of these twelve oft-cited assertions is, then, dubious at best. Those who uncritically accept their validity are less likely to help make the world safer from the threat of nuclear war than those who question them. So while we doubt that these "givens" are relevant to government policy, they are less irrelevant than the first set of commonly held assumptions.

How Is Nuclear War Different?

The misconceptions having to do with nuclear war include more than these twenty-four variations on the theme of apocalypse. Indeed, many people are convinced that the advent of nuclear weapons—the nuclear era—has dramatically changed the world in general and military strategy, doctrine, and tactics in particular. There is no question that enormous changes have in fact occurred, the most obvious being the change in the rapidity and magnitude of potential destruction. But there are probably as many dramatic continuities as changes. First, the differences.

The most outstanding difference between the nuclear era and previous millenia is the terrifying threat of the instantaneous and total destruction of whole countries, if not of mankind itself. Many people are convinced that a nuclear war would be the end of human history, and there is a pervasive fear of fallout, thermal radiation, genetic defects, and environmental chaos. The conventional wisdom simply expects a nuclear war—any nuclear war—to be a literal Armageddon.

Antinuclear protesters and sympathizers warn of inevitable destruction; they argue that any preparation for fighting and surviving a war is stupid, dangerous, and/or immoral. Issues that were once appropriate military concerns have become discredited perversities; it has become almost obscene to note that a nuclear war might be limited, or that one side might win.

Their point is not without some merit. The total number of military and civilian fatalities in World War II reached an estimated 50 million, but that was over a long period of time and among many nations. The casualty level in even a limited U.S.-Soviet nuclear conflict could conceivably reach 50 million within a few hours. (It could also be a small fraction of that number.)

In addition, there is a whole new fear of accidental war caused by mechanical or human error. In the prenuclear era, this was simply not a realistic possibility. Even if there were misunderstandings or crossed signals, no immediate disaster would result, except perhaps in a very tense crisis. Today it could, and this perception has led to some universal changes in national behavior.

Primarily, there has been a unique degree of restraint among nations. The nightmare of an unchecked spread of nuclear weapons is less frightening than it once was, mostly because proliferation has proceeded much more slowly than was feared twenty years ago. Most countries have not raced to acquire nuclear capabilities, and those that have made the effort have encountered greater difficulties than they imagined. Voluntary, unilateral limits on the acquisition, deployment, and role of nuclear weapons have been observed by a surprising number of nuclear and "almost-nuclear" powers. Self-imposed political, moral, and psychological restraints have been more important than international, legal, and coercive measures. Voluntary nonproliferation policies, more than the Nonproliferation Treaty of 1968, have worked surprisingly well. In the past, very few countries would have looked at the dynamics of escalation and voluntarily forgone the highest-quality weapons they could obtain. Now a number of nations capable of developing nuclear weapons have eschewed this option.

The persistent fear of nuclear disaster, then, has brought about some positive consequences: the potential for tremendous damage and suffering has provided national leaders with almost overwhelming incentives to be careful in their conduct of international politics. A relatively favorable climate for arms control has also been created—for both explicit and implicit agreements—although the difficulties of achieving and maintaining reliable and useful arms-control measures remain at all levels.

In *The Absolute Weapon,* the first great work on military strategy in the nuclear age, Bernard Brodie wrote: "Thus far the chief purpose of our military establishment has been to win wars. From now on its chief purpose must be to avert them." * This observation is, of course, exactly right. The notion of Armageddon has created a self-defeating prophecy: the more frightened decision makers are, the more careful they become and the less likely they are to initiate any action that might bring about the dreaded escalation to nuclear use. Armageddon is averted to the degree it is feared. As a result, nuclear deterrence has not only worked remarkably well in peacetime and served to limit the scope and intensity of conventional warfare, but it is likely to work surprisingly well in a large U.S.–Soviet conventional, and even a limited nuclear, war.

One significant indication of the effectiveness of deterrence is that the Soviet Union and the United States share the belief that a nuclear war would begin only out of desperation or inadvertence. A weakening of deterrence could increase the possibility of a calculated "voluntary" war or worse.

Deterrence is part objective and part subjective. The objective part is that adequate military preparedness could be critical in some important circumstances. The subjective is, in part, the *perception* by the potential enemy of the other side's power, determination, and courage, as well as its estimate of the wartime effectiveness of the other side's relative nuclear capability. If the aggressor believes that his opponent

* Bernard Brodie, ed., *The Absolute Weapon: Atomic Power and World Order* (New York: Harcourt, Brace & Co., 1946), p. 76.

has the weapons and will to use them, deterrence is likely to be effective. Therefore, *a nation's defense lies, in part, in the credibility of its threat to use nuclear force as a last resort,* which means that the need remains for coherent and plausible policies for the use of nuclear weapons. Maintaining this as an option has meant a major change in approaches to the problems of war.

How Is Nuclear War the Same?
(or, How Much Has Not Changed So Greatly?)

The existence of nuclear weapons has therefore changed the nature of warfare greatly, but it has not made warfare totally obsolete. The primary continuity between the prenuclear and nuclear eras is that wars can still be fought, and there is still a need to plan and prepare for fighting and even winning a war if deterrence fails.

Many other classic political, strategic, and military concepts remain valid, although the advent of nuclear weapons has brought about additional formulations as well (e.g., "deterrence only"). But a complete reliance on deterrence-only policies risks too much and discounts many of the continuities with the prenuclear era. Primarily, it risks the almost-certain destruction of both U.S. and Soviet cities in the event of even a single war-prone misadventure. An informed and conscious decision to maintain a peacetime military posture that is incapable of anything other than destroying the enemy's civilian population and industry is both immoral and irresponsible, vicious as well as mad. It also risks either blackmail by an aggressor who assumes that the United States would rather surrender than fight a countercity nuclear war, or another "Pearl Harbor"—in which case, despite (or because of) its minimum nuclear force needed for deterrence only, the United States could suffer a dramatic and overwhelming initial defeat. Alternatively, deterrence-only policies could result in a conventional defeat of the United States that would be much less than Armageddon, but drastic nonetheless.

One of the most important continuities of the nuclear era is the fact that war can still be fought, terminated, and thought of as an experience to be survived rather than as the end of human history. As noted above, there is no reasonable evidence suggesting that a nuclear war would leave the world lifeless. Rather, there would be far more survivors than fatalities, by factors ranging from five (about a billion dead) to a thousand times that. Most people find numbers like this clearly unacceptable, which they are, but the range of uncertainty is literally that great —and the thought that there would still be an extraordinary number of survivors should be perversely comforting. More important, we should be concerned with making the fate of the survivors a more tolerable one.

As in conventional war, a great deal will depend on how a nuclear war is fought, what preparations are made to minimize damage, and how extensive the conflict is. Like past wars, nuclear wars could come in different "shapes" and sizes, and would require different resources and defenses—and have quite different outcomes.

But the problems of war fighting, war termination, postwar recovery, and the long-term consequences of war are complicated by nuclear war's unprecedented nature. These issues still must be addressed by responsible national leaders, who must still be actively committed to ensuring that their countries endure even an "unthinkable" experience. The unknown can make many of these issues very difficult to deal with, but as far as we know, they need not be overwhelming obstacles to continuing security and survival.

A nuclear war would probably not come as a result of an attack out of the blue; it might well begin with an old-fashioned declaration of war. Almost all defense analysts believe that an unexpected attack is highly unlikely—instead, a conflict probably would result from a period of intense crisis. A declaration of war might still be of considerable value: it would make clear the extreme seriousness of the situation without necessarily producing or requiring an immediate resort to armed conflict. Or it might lead, as in the beginning of World War II, to a very limited armed conflict followed by a lengthy period of "phony war." A declaration of war followed by a phony war probably would prevent an immediate de-escalation of the crisis, but might lead the aggressor to back down eventually. It would give both sides time to mobilize—to work on making themselves stronger instead of fighting with each other.

This concept of mobilization—gearing up a nation's economy and society to protect and defend its people—is another notion that precedes, and yet has applications in, the nuclear era. The U.S. military buildup after the outbreak of the Korean War is a perfect example of how effective a mobilization can be. Congress was debating whether to increase the nation's defense budget from $13 billion to $14–16 billion (by 8 to 23 percent) when North Korea invaded South Korea; Congress immediately authorized the Department of Defense to spend up to $60 billion (an increase of 350 percent). This and subsequent authorizations resulted in U.S. military superiority over the Soviets for the next two decades. Many analysts, including myself, believe that the Soviets fear a new U.S. mobilization more than they fear a direct U.S. attack.

Mobilization involves the massive diversion of civilian resources and production facilities to military needs. It proceeds with wartime urgency and priority, but occurs before any major destruction of cities or economic infrastructure has taken place. In many scenarios, mobilization could be decisive, even without much fighting. Urban evacuation plans, fallout and blast shelters, and other preparations for survival and postwar recuperation could decrease enormously the vulnerability of civilians to the effects of nuclear war, and thereby increase the ability of both sides to bargain effectively in a serious crisis, as well as reduce the frequency of such crises.

Mobilization can be dangerous, but usually less so than going to war. In most instances it is a way of de-escalating an intense war-prone crisis: if it does not lead to a settlement, it can still make the difference between a low or high level of hostilities and between a relatively "successful" war outcome and total disaster. (High officials in the Reagan administration fully recognize the importance of mobilization capabilities, but have run into a variety of obstacles in attempting to develop them.)

In the nuclear age, war can still occur and this possibility must still be taken into account, even though we assign plans and preparations for fighting a nuclear war a lower priority than deterrence. We must continue to have a variety of credible alternatives to "peace at any price," even as we strive more than ever not to resort to them.

Some Useful Terms

The similarities between prenuclear conflict and nuclear war, therefore, include many political, military, and conceptual aspects, even as the differences include many new technological and psychological factors. In addition, the existence of nuclear weapons has engendered a whole new vocabulary, some of which is fairly simple and enlightening. But many of the terms and phrases have taken on such emotional bias that they are sometimes less than useful (and often misleading) in the public debate on nuclear war issues.

I would argue that most of the vocabulary should be retained, since it has by now become familiar and relatively acceptable, but it is important to clarify both the denotation and connotation of some of these terms.

Deterrence Only. The word "deterrence" comes from the Latin *deterrere*, "to turn aside or discourage through fear"—hence to prevent some nation, group, or individual from doing something as much through sheer terror as through rational, "cost-effective" analysis of the consequences. Many of the policies supported by members of the antinuclear groups are based on deterrence only. (Some policies of the U.S. government are also partially based on deterrence only.) In one typical policy, mutual assured destruction (MAD), the premise is that if either side attacks the other, both would be totally destroyed or severely damaged; knowing this, the potential attacker would not attack. Under MAD, the total military commitment is to dissuasion through terror.

MAD is the most straightforward of the deterrence-only policies. Supporters generally advocate adequate provisions to prevent accidental and other inadvertent war, survivable nuclear forces capable of responding to a reliable "go-ahead" order, and the targeting of nuclear weapons solely and inevitably against the opponent's urban-industrial complex. They consider all other nuclear war planning or weapon systems to be unnecessary or counterproductive. For example, they support no counterforce capabilities or strategic defenses that might strengthen deterrence or alleviate the consequences of a failure of deterrence. Hence MAD is a deterrence-only posture.

The MAD policy fits its acronym: it is somewhat insane (and immoral) to suggest that a nation should risk suicide in pursuit of peace. No one argues with the objective (despite some uncertainty about what is meant by "peace"), but sole reliance on deterrence is deceptively comforting—so uncomplicated, inexpensive, convenient, and politically and emotionally acceptable that it seems practical and even moral. In fact, it is gravely defective—politically and militarily, and especially morally.

Finite and Minimum Deterrence. These are variations on deterrence only. Advocates of "finite deterrence" imply that their opponents want an "infinite" (excessive) deterrence strategy. They support a "reasonable" and "moderate" deterrence-only strategy. Often they support a "minimum-deterrence" position, which holds that one does not need many nuclear weapons to deter a nuclear-armed opponent. Only a minimal nuclear retaliatory capability is required, since the overwhelming

damage that could be wreaked by even a few nuclear-weapon detonations would be enough to make any enemy calculations of surviving a
war very questionable. Proponents of finite and minimum deterrence
criticize any additional nuclear forces as overkill. In line with this reasoning, MIT Professor Jerome Wiesner has argued that a unilateral U.S.
moratorium on the building of new nuclear-weapon systems would be
"safe," because deterrence requires "not the tens of thousands of
bombs in existence, but the certain ability to deliver 200 on either
side." *

Reinforced Deterrence. Many of those who support deterrence-
only as an objective nonetheless disagree that dependence on the threat
of assured destruction of the opponent's urban-industrial complex
should be the sole means of achieving that primary objective, and also
argue that it is reckless to guarantee (by inadvertence or design) that
even an accidental war would unfold into an unmitigated catastrophe.
Further, they argue that Soviet decision makers are as much (if not
more) deterred by fear of the destruction of Soviet military power or
the loss of political control resulting from a nuclear war, as of large-
scale damage to the Soviet population and industrial base. The ability
to deny victory to the Soviet Union, or to destroy the Soviets' most
valued political and military assets, they say, might be more of a deterrent than a simple MAD policy. Advocates of this approach favor "reinforced deterrence."

Deterrence Plus Insurance. Other supporters of nuclear deterrence
(but not deterrence only) opt for "deterrence-plus-insurance" policies.
They recognize that deterrence can fail and feel a nation must be prepared to lessen the consequences of such failure by trying to obtain an
"improved war outcome." Improved war outcome denotes the strategic
aim of limiting war damage suffered by the population and resources of
one's own nation and its allies, and of improving, so far as possible, the
military and political outcome of a war.
   Many deterrence-only proponents reject any "insurance" against
nuclear war, arguing that it might lull political decision makers, the
military, or the populace into believing that their country could survive
a war, and might therefore promote reckless attitudes. Even worse

* "Is a Moratorium Safe?" *Bulletin of the Atomic Scientists,* August-September 1982,
p. 6.

(according to these critics), the enemy might believe that measures for an improved war outcome imply a greater willingness to risk war, and this might lead him to increase his own offensive forces, and perhaps to preempt in a crisis because he could not be sure the other side would be deterred from striking first. Thus, they argue that anything more than deterrence only is likely to create or accelerate an arms race or otherwise make it more dangerous.

As I mentioned above (and will argue at greater length later), the prospective human suffering and material devastation for even a small nuclear war is so great that the likelihood of a limited nuclear war is unlikely to diminish by very much the circumspection with which both U.S. and Soviet leaders approach the prospect of such a conflict. Given that the Soviet Union has already invested heavily in strategic offensive forces and defensive programs consistent with the objectives of improved war outcome, the alleged action-reaction effects of the deterrence-plus-insurance posture are undoubtedly exaggerated. To the extent damage-limiting policies and programs did increase the probability of war, this slightly greater risk would be outweighed by the insurance against the consequences of nuclear war afforded by active and passive defensive systems and counterforce weapons.

War Fighting. The deterrence-only positions stand in contrast to what is sometimes called a "war-fighting" posture. War fighting does not advocate war: it simply means the nation should be concerned with how a war will be fought and how it will end if deterrence fails. Limitations on how nuclear forces are used (such as prohibition of counterpopulation attacks) are typical of what professionals mean by the term war fighting. Any interest in the wartime employment of nuclear forces and in the ways in which a war might be brought to an end short of mutual annihilation constitutes a concern with war-fighting strategies. It is incorrect and unproductive to categorically accuse those who subscribe to war-fighting concepts either of wanting to fight a nuclear war or of having less interest in deterrence.

It is true that some advocates of war fighting may pursue objectives that exceed deterrence and insurance. As with deterrence only, war fighting covers a range of different positions related to nuclear targeting and force requirements. Proponents might argue for war fighting from a purely prudential point of view (in which case it is essentially a deterrence-plus-insurance stance), or could advocate it as a way of denying victory to the enemy (essentially for reinforced deterrence), or as a means of obtaining foreign policy advantages (essentially to support "extended deterrence," as described below).

If strategists who favor war-fighting plans and capabilities try to achieve greater flexibility in foreign policy through implicit or explicit nuclear threats (or use), then they are attempting to improve the nation's "extended deterrence." Extended deterrence refers to the use of nuclear forces to prevent the enemy not only from attacking one's own homeland or major military forces, but also from attacking (or threatening) one's friends and allies, or neutrals. Extended deterrence can improve the effectiveness of deterrence only by decreasing the number of occasions on which such deterrence is likely to be strained—i.e., by deterring provocation or making an opponent more willing to settle an escalation-prone crisis.

The use of nuclear threats to protect one's allies raises a question: can such threats usefully and credibly help preserve the peace generally, or only prevent an attack on one's own country and forces? If it is the latter, what happens to one's allies? At present, the United States probably would respond much more forcefully and reliably to an attack on itself than on its allies. Yet Britain went to war in 1914 and 1939 because its allies were attacked, not because Britain itself was. Thus, it is important to maintain an option to go to war even if one is gravely provoked but not attacked directly.

Nuclear Weapons, Morality, and the Future

The current nuclear debate requires not only a new lexicon, but also an examination of the moral questions raised by the existence of nuclear weapons and the dangers of nuclear war. Each side in the debate has referred to moral considerations in making its arguments.

For example, President Reagan, in an address to the National Association of Evangelicals, told his listeners that in taking a position on the nuclear freeze they should

> beware . . . the temptation of blithely declaring yourselves above it all and label both the United States and the Soviet Union equally at fault, to ignore the facts of history and the aggressive impulses of an evil empire, to simply call the arms race a giant misunderstanding and thereby remove yourself from the struggle between right and wrong and good and evil.

I ask you to resist the attempts of those who would have you withhold your support for our efforts . . . to keep America strong and free, while we negotiate real and verifiable reductions in the world's nuclear arsenals and one day, with God's help, their total elimination.

The day of the president's speech, a different group of religious leaders declared the federal budget to be the government's most important "moral statement," and said they were "profoundly disturbed by the vision which emerges from the fiscal 1984 edition of our statement of moral purpose. It is a vision of a nation intent on a selfish and dangerous course of social stinginess and military overkill."

In the United States, it is highly unusual for moral questions to be linked to policy issues with blatant, almost gospel-like fervor. Americans tend to deal with policy matters in relatively practical and humanistic terms—not usually as moral issues, and even less as religious ones.

And yet, since the time of the Puritans, Americans have always had a strong sense of what is morally acceptable, even though in most instances pragmatism has held sway. During the post-World War II era, most Americans have accepted nuclear weapons as facts of contemporary life to which moral judgments do not apply—or at least where morality takes a backseat to national security.

But many concerned individuals argue that with regard to nuclear-weapon policies, the traditional military, political, and national-security concerns are becoming increasingly irrelevant. They question whether the defense of one's country is either honorable or noble. Some argue that the only thing that matters is to end the "arms race" and thus reverse trends that make nuclear disaster inevitable. This "apocalyptic premise" is becoming more and more prevalent. From this perspective, morality and pragmatism are the same thing. The nuclear freeze movement is partially a popular manifestation of this position.

The recent involvement of religious leaders in the nuclear debate is perfectly reasonable. The moral and spiritual aspects of nuclear war are their legitimate concerns. But can the concepts of good and evil determine the appropriate size of our nuclear forces, or whether this country needs policies other than just MAD, or whether a nuclear freeze would make the world more or less safe?

Judaism and Christianity have never required their adherents to be saints or pacifists, but merely to pursue the ends of peace and justice. The traditional Church criteria for the justification of violence are set forth in the doctrine of the just war. This precludes war except in the pursuit of "peace and reconciliation," and justifies war only if it is waged to secure "basic rights," to promote a "decent human existence" or to protect the innocent and righteous. The benefits of violence must be "proportionate" to the human and other costs, and one must be able to discriminate between actions against an aggressor (which are justified) and those that hurt noncombatants (which are not).

The National Conference of Catholic Bishops concluded that nuclear war is beyond any useful application of the just-war theory, primarily because the principles of proportionality and discrimination would be virtually impossible to observe. According to the bishops, a nuclear attack that wipes out a city is automatically disproportionate to any provocation, and an attack restricted to military targets that nonetheless harms millions of civilians is indiscriminate.

While antinuclear arguments based on moral interpretations are becoming more widely held and are playing an increasing role in the nuclear debate (in some cases dominating it), I will argue that many of these ideas are incorrect, misperceived, misleading, or incomplete, and will become even more so in the future. The main reason for public acceptance of flawed moral arguments (e.g., some of the reasoning in the Pastoral Letter) and unwise policy prescriptions (e.g., the nuclear freeze proposal) seems to be a general lack of compelling alternative concepts and proposals put forth by the government.

Consequently, it is important that the government (and defense experts in general) begin now to fashion, implement, and explain a different approach to the control of nuclear weapons, one that incorporates each of the following considerations:

1. A long-term perspective
2. Political and moral imperatives
3. Optimistic political and military "visions of the future"
4. The promise of future technological advances

A long-term perspective is necessary because nuclear weapons cannot be disinvented (total worldwide disarmament was the second of the twelve nonissues). In this sense, the dangers associated with these weapons constitute what might be termed a "fifty-year" problem. Because the problem cannot be eliminated, it must be controlled. In my view, "control" involves preventing the use of nuclear weapons altogether, decreasing their influence in international affairs, and creating limitations on their use (and effects) if nuclear deterrence breaks down.

By taking a long-term approach to nuclear-weapons control, we can avoid being stampeded into the pursuit of chimerical or dangerous "solutions." Because of Jonathan Schell's antinuclear alarmism—"at any moment our lives may be taken away from us and our world blasted to dust"—he urges on his readers both chimerical (the redemption of mankind) and dangerous (a nuclear freeze) means of coping with the threat of nuclear war. Although one cannot deny that someday nuclear weapons might be used, the outbreak of nuclear war is not comparable to the unexpected occurrence of a cataclysmic natural disaster (nor is it necessarily an Armaggedon). We have adequate time to formulate prudent policies, as well as time for favorable technological, economic, and political developments to unfold that will promote *greater* long-run safety in a nuclear-armed world.

But the need for a long-term perspective does not mean that we should relax in our current efforts to preserve peace and security. Proponents of arms-control negotiations with the Soviets are often willing to admit that the resulting agreements are far from satisfactory, but argue that with the negotiations, "We're trying to buy time." For over twenty years my response has always been, "Yes, but what is your plan for how to use that time?" (I rarely get an answer to my question.) Similarly, the authors of the Pastoral Letter are right in asking that the time afforded by nuclear deterrence be used to create "a long-term basis for peace" (although I would disagree with their apparent willingness to abandon deterrence altogether, and their emphasis on "progressive disarmament").

This "long-term basis for peace" should be responsive to the political and moral imperatives that we see today and are likely to endure. One of the most significant lessons of the current nuclear debate is that decision makers can no longer assume that the public will be indifferent or diffident toward government defense and arms-control policies regarding nuclear weapons. Worldwide economic development will continue to fuel "social-limits-to-growth" movements (like the "Greens" in West Germany and many profreeze groups in this country), with their anti-industrial and antinuclear attitudes. Defense policies and programs in the U.S. and Western Europe require sustained public support over the long haul, because—in addition to the "fifty-year problem"—most weapon systems require ten to fifteen years from conception to deployment, and recent arms-control agreements (SALT II) have taken nearly a decade to negotiate. Disruptions caused by public opposition need to be avoided. Providing U.S. (and NATO) defense strategy with a firm moral underpinning is one way of taking into account growing antinuclear sentiments without compromising the security of the United States and its allies.

In this respect, the defense establishment should view the Pastoral Letter and the moral discussion it has aroused as a terribly important opportunity and not as a minor irritant. By endorsing and adopting two of the moral judgments rendered in the letter—"no first use" of nuclear weapons and no deliberate nuclear targeting of civilian populations—the government would go a long way toward seizing the moral as well as the political and strategic high ground.

For a long time I have favored no first use. By no first use, I mean that the only legitimate functions of a nuclear arsenal should be to: (1) *deter* the outbreak of nuclear war; (2) *counterbalance* the possession of nuclear weapons by others; and (3) *correct* for the use of nuclear weapons by others. While I support a U.S. no first use pledge (which both the Soviets and the Chinese have announced), I part company with the Catholic bishops (and some other prominent advocates of no first use) by urging that a reinforcement of NATO's conventional forces and a strengthening of U.S. strategic forces accompany such a declaration.

Improvements in NATO's conventional-force posture would compensate for a withdrawal of the current U.S. threat of a first strike to defend NATO against a conventional attack by the Soviet Union and its East European allies. Bolstering U.S. strategic forces with new systems (like the MX ICBM, the Trident ballistic missile submarine, the B-1B bomber, and air-launched cruise missiles) would allow the United States to deter *Soviet* first use (by a more credible threat of second use), which, if unchecked, could negate any increases in the effectiveness of NATO conventional defenses.

In addition, contrary to the suppositions of the bishops, strategic forces with higher-accuracy, lower-yield nuclear warheads and better command-and-control systems would be needed to avoid unwanted civilian casualties that could result from attacks limited to military targets. Strategic defenses similarly could save the lives of tens of millions of noncombatants and preserve a base for postattack recuperation.

Even though our first priority is to maintain nuclear deterrence, we have a moral obligation and the right of self-defense to hedge against its breakdown. In contemplating the failure of deterrence, we must consider what plans and actions would best satisfy the most appropriate moral, political, and military criteria for conducting a war. Which strategies, tactics, and military and nonmilitary systems would best protect citizens (U.S. and Soviet) and limit the scope and intensity of the conflict? The bishops' moral stricture against "counterpopulation warfare" is certainly one of the most important criteria,* but they are wrong in their practical judgments opposing the means—i.e., war-fighting forces, planning for limited nuclear war, and civil-defense programs—needed to adhere to the principle. Moreover, ongoing developments in weapon technology may yield long-range nuclear (and perhaps more important, nonnuclear) forces that are both "proportionate" and "discriminating" in the sense of the just-war doctrine: proportionate, in that the damage they do is far less overwhelming than the military and political functions they perform; and discriminate, in that collateral damage to the enemy's civilian population can be largely avoided.

Besides assuming a long-term perspective and acknowledging the positive aspects of certain political and moral imperatives, the government should adopt and offer the American people at least one optimistic "vision of the future" with regard to nuclear weapons. (Sometimes I have referred to this notion as a "happy ending to the arms race.") To the extent that there is today a popular image of the future course of the nuclear-arms competition, it is by and large an uninspiring one of "business as usual," an unrealistic one of total disarmament, or a terrifying one of an ever-spiraling arms race that almost inevitably leads to catastrophe. Neither the first nor the third of these pessimistic "visions" strikes me as very plausible, though both contain elements of what might unfold in the future. The third is especially troublesome, since it engenders undirected fear that in turn could lead to widespread despair and paralysis that would undercut support for the defense measures necessary to prevent nuclear war *and* deter Soviet aggression.

An alternate vision of the future does not have to be a prediction of what will happen, but only a "surprise-free" scenario—i.e., a projection that would not cause astonishment if it became reality. An appropriate vision of the future can be a set of conditions that we work toward bringing about, as much as a scenario that we believe will come to pass. In particular, I believe that the emergence of an inherently more stable multipolar world is likely to occur.

The post-World War II configuration of two superpowers is eroding.

---

* It should be noted that I would modify somewhat the letter's prohibition of counterpopulation warfare to read, "No *routine* targeting of civilians." As a strategist, I can conceive of nuclear war scenarios in which the United States would need to threaten (but not necessarily undertake) the destruction of Soviet cities in order to deter Soviet attacks against American civilians.

I believe that early in the twenty-first century there will be around eight major powers, each of which will have a gross national product of about $1 trillion or more (in 1980 dollars) and all of which will have the technological capability to "go nuclear" in a serious way, even if they do not actually acquire nuclear weapons. This kind of multipolarity, though many fear it, seems more likely to make the world safer than not, even with some potentially difficult problems of transition (the nuclear armament of Japan and West Germany, for example).

In a multipolar world of several nuclear-armed Great Powers, the victor in any war between two nations could not expect to dominate the world automatically; it could not even expect to maintain its prewar preeminence among nations, since resorting to war might involve great damage to itself (even if it won) and would be a basically unacceptable form of political behavior. Multipolarity would also mean a greater reliance on balance-of-power politics, an international system that worked remarkably well for almost a hundred years in Europe (from the Congress of Vienna to the outbreak of World War I). Within our projected group of eight major world powers, shifting alliances and pressures could help maintain a fundamental equilibrium that is simply not possible in a two-power world.

Perhaps the most reassuring result of this multipolarity would be the strong impetus it would give to the adoption of more or less sensible nuclear postures by the United States and other countries. Widely held (but irresponsible) policies such as MAD and "launch on warning" * would become manifestly reckless. Policies and programs that would be more appropriate would include improvements in planning and forces for waging "controlled" nuclear war, the buildup of active and passive defenses, and the acceptance of no first use as part of a worldwide antinuclear taboo.

The greater stability of a multipolar world is one possible vision of

---

* "Launch on warning" presently means that U.S. strategic forces, particularly land-based ICBMs, would be launched once it became clear that the Soviet Union had initiated an attack on the United States. Thus, U.S. missiles would be launched before Soviet warheads could destroy them. Such a policy would mean that the Soviets could not destroy U.S. ICBMs in a preemptive attack. However, this policy has many drawbacks. It is almost certainly more accident-prone than the current system. A decision to launch would have to be made very quickly after reports were received from our distant early warning system that a Soviet strike was on the way. If this information proved incorrect, the ICBMs that had been launched in reply could not be called back. The result could be a nuclear war that would otherwise have been avoidable. The dangers associated with a launch-on-warning posture would be especially acute in a tense crisis in which war was feared and strategic forces had been moved to a higher state of alert.

the future that would gradually evolve as the result of political, economic, and military trends. Alternate visions can also be the result of deliberate government policies. In a well-known article published many years ago, George Kennan sketched a more or less optimistic outcome for the U.S.-Soviet competition. He argued that through the application of a combination of military and diplomatic "counter-pressures," the United States could stymie Soviet aggression, and this "patient but firm and vigilant containment" would "promote tendencies which must eventually find their outlet in either the break-up or the gradual mellowing of Soviet power."* While the effectiveness of containment in tempering Soviet aggressive tendencies is debatable, the point here is that Kennan articulated a strategy for the long run that had a positive payoff and thus was worth working toward.

In March 1983, President Reagan delivered a watershed address that offered another "vision of the future" (a term he used in the speech) based on the development of ballistic missile defenses (BMD). The president's strategy not only is optimistic, long-term, and responsive to political and moral imperatives, it also calls for the exploitation of new technologies (e.g., various kinds of lasers and particle beams). BMD, coupled with arms restraints, can help increase safety and security through a strategic posture we studied at the Hudson Institute in the early 1960s: "Arms Control Through Defense" (ACD)—or "Defense Through Arms Control." With ACD, arms control (e.g., deep reductions in nuclear forces) can be made to work because the deployment of active and passive defenses permits greater trust of controls on strategic forces, and defenses can be made to work better because of the limitations on strategic offensive forces. Furthermore, extensive strategic defenses, by decreasing the striking power (and thus the threat value and prestige) of small nuclear forces, can make the acquisition of nuclear weapons a less attractive option for nonnuclear countries, thereby aiding nonproliferation policies. And if a nuclear war does break out, strategic defenses (along with doctrines and capabilities for controlled war) make it much less likely that the outcome of the conflict would be "mutual homicide."

The appearance of long-range, superaccurate missiles, carrying *nonnuclear* warheads (or very low-yield nuclear warheads) is another potential technological development in the U.S.-Soviet strategic competition that most likely would have favorable consequences. Since

* "The Sources of Soviet Conduct," *Foreign Affairs*, July 1947, pp. 566–82.

their advent in the late 1950s, ICBMs (and later SLBMs) have undergone very impressive increases in their delivery accuracies. The MX ICBM should be able to deliver its warheads to within a few hundred feet of targets in the Soviet Union. The Pershing II intermediate-range ballistic missile (with its maneuvering reentry vehicle) and the long-range cruise missiles, in certain senses, are the forerunners of the kind of systems I am referring to here. With such weapons, keeping a nuclear war limited would be a considerably more feasible task. One could imagine a central nuclear war* being fought in which cities and other heavily populated areas went largely unscathed. Such precision bombing would facilitate the observance of the just-war principles of proportionality and discrimination.

Some will object that "the people" rather than the defense establishment should determine the future course of our nuclear weapons policies. Antinuclear activists often, but inaccurately and unfairly, portray professionals who study (and prepare for) war—in order to prevent it— as a kind of sinister elite. Yet soldiers are sometimes more profound about war than philosophers; generals can sometimes articulate national purposes better than statesmen. Similarly, military planners and nuclear strategists ought not to be discredited out of hand. Many of them understand better than most the "immorality" of nuclear war—including *not* preparing for it.

This is not to say that civilian defense planners and decision makers or senior military officers are necessarily wise and judicious, but only that many feel their responsibilities heavily and that their special knowledge and expertise—as well as their special concern—should not be automatically impugned or ignored as that of a special-interest group or evil and bloodthirsty individuals bent on destruction. Neither, however, should one look to military experts for salvation. In this field more than in most, there is no "bottom line" (i.e., no genuine reality-testing short of war or intense crisis); as a result, many defense experts are as prone to passions, illusions, and emotional judgments as anyone else. Their positions are inherently no more and no less accurate, objective, moral or immoral than the views of laymen, only sometimes more sophisticated (and sometimes more intimidating).

Those advocating a freeze, unilateral disarmament, or some form of pacifism do not have a monopoly on morality when it comes to nuclear

---

* "Central nuclear war" refers to a war between the United States and the Soviet Union in which nuclear weapons are used against the heartland and/or major strategic forces of both powers.

weapons. Our common moral obligations are to: (1) preserve nuclear deterrence on terms consistent with our security interests; (2) improve the safety of the world in the face of the dangers posed by nuclear weapons; and (3) alleviate the consequences of a nuclear war if our best efforts to prevent such a conflict fail. Plans for the long haul, a prudent no-first-use policy, visions of the future, and strategic defenses (among other things) are important initiatives to further these objectives. Vigilance and appropriate defense efforts constitute the best way to avoid war and insure safety in the future. Here, as elsewhere, there are no panaceas, no free lunches, no ultimate and complete solutions. But determined efforts by large numbers of responsible people can make a difference. It's time we got started.

2

How to Think About the Unthinkable: Some *Gedanken* Experiments

Herman Kahn was well aware of his Dr. Strangelove image among many (basically hostile) nonmilitary audiences, yet he was often asked to address them on defense issues. He found that one useful way to break the ice was to begin by taking a poll.

—How many people in this room believe that the number of nuclear weapons can be reduced? Almost everybody believed they could be.

—How many think they can be reduced to zero? Almost no one thought so; they recognized the fact that nuclear weapons would continue to exist.

—How many people think the United States should unilaterally reduce its nuclear arsenal? Most of the audience felt this was not the best option.

—How many of you believe arms reduction should take place through negotiated treaties like SALT or START or a mutual freeze? Most people supported these as the best alternatives.

—How many of you believe that the remaining weapons might actually be used? Most people feared this as a very real possibility.

"Now that we are agreed that nuclear weapons might be used," Herman would continue, we have to think about how to use them: against what targets? Toward which objectives? We have to think about how to use them to maximize the chances of survival and minimize the damage (to civilians, military installations, industrial complexes, and so on). And we have to think about how to use them to end the war as soon as possible.

Herman would stress the point that it's not immoral to think about these things; on the contrary—planning for nuclear contingencies in the event deterrence fails is the only responsible, prudent, and moral course. The planners, Herman would say, don't plan war as a deliberate act; planning is a prudential exercise, like taking out insurance against an accident. You hope you don't need it, but you've taken reasonable precautions just in case.

By the end of this dialogue, thinking about the unthinkable had usually ceased to be an obscene act.

To help audiences come to grips emotionally with some of the concepts of thermonuclear war, Herman used questions and answers, anecdotal illustrations, and "Gedanken" experiments. Gedanken is the German word for "thought." A Gedanken experiment is a well-known analytical tool in philosophy and physics; its purpose is to think through what would happen if a certain experiment (in accordance with the physical laws of nature) were carried out, even though it would probably never actually be performed. Nevertheless, the process of thinking through the experiment can be very helpful in understanding certain issues (although it can also be misleading if the assumptions of the experiment are fallacious or the conclusions misapplied).

The following are some examples of Herman's descriptive lectures.

The fact that no thermonuclear war has ever been fought means that in one sense, everybody, including the "experts," is an amateur. If there were a nuclear war it might turn out (particularly if the war were short and not complicated by unexpected actions of the enemy) that a well-prepared attack might unfold more or less according to plan, and the results would be reasonably predictable. But no one can have any justifiable confidence that this would happen, even if the war plans were executed very smoothly—and few believe that all the plans would go very smoothly in a nuclear war. Enemy actions and various operational and situational uncertainties could complicate or even derail well-prepared plans. Even more unpredictable would be the possible aftereffects of the war: political, military, environmental, economic, medical, and so on. The importance of the unknown, which must always be kept in mind when thinking about or discussing nuclear war, is made clear in the following conversation (perhaps apocryphal) between a senior military officer and a young Rand civilian defense analyst.

The officer asked, "How can you, a young man who has never fired a gun in anger, have the temerity to argue with me on nuclear war, on its strategy and tactics?" The young analyst replied:

> Our studies indicate that it takes about ten nuclear wars to get a sense of the range of possibilities—indeed, this is a very minimal level of experience. Just out of curiosity, how many such wars have you actually fought or even studied? I might also point out that much of the experience gained in conventional warfare appears to be misleading when applied to nuclear war problems.

The general admitted that he had no such experience or background. The analyst then continued:

> Why don't we two amateurs get together and see if we can't work out some of the issues concerning nuclear war? It's clear that you know a lot more about the practical aspects of *peacetime* military equipment, operations, training, field exercises, morale, interservice politics, and so on, but I rather suspect I've thought longer and perhaps more deeply about other questions regarding nuclear war that don't come up in peacetime.
>
> If fact, you are kept increasingly busy working on peacetime problems, many of which are "urgent, but not necessarily important." In contrast, I tend to spend more time and interest on the "important, but not necessarily urgent" issues, including the questions of how a war might start, how it should be fought, how it might be terminated, and what aftereffects it might create. But when it comes to real wartime strategies and tactics, we are both almost complete amateurs.

One could sympathize with the senior officer at the brashness of this young man, who may well have had illusions about the relevance or value of some of his ruminations and analyses. Nevertheless, the analyst did show a modicum of humility by insisting that both were amateurs, an acknowledgment that is absolutely correct as far as "hands-on experience" or the study of relevant historical examples is concerned.

This lack of either first or secondhand experience is central and disturbing. While both the officer and the analyst may have useful background or understanding, there are presumably equally important kinds of knowledge and experience which they (and everybody else) lack. In thinking about nuclear war, we are trying to get a handle on something which is not only quite complex, but is also quite theoretical. We are not able to test adequately (or at all) many of our ideas about nuclear war. No nation, political leader, or military staff has undergone a conflict that escalated to nuclear war. Untested concepts in any area may turn out to be false or otherwise flawed when they are actualized; but with regard to nuclear war, this is true to a degree that is terrifying. In the past, many innovators, in setting out on their explorations, were uncertain about what faced them, but the consequences of error were never so great.

Let us start with an examination of a crude political context for nuclear war. This may betray a lack of imagination or a desperate attempt to fall back on historical examples because these are all we have, but analogies with simplified situations can be drawn. The practices of certain traditional or primitive societies can be instructive as to what might happen in a nuclear confrontation or a nuclear war. Current international relations share many interesting and important characteristics with certain primitive societies. Neither system is anarchic, but both have elements of anarchy in that members maintain the right to be the judge of their own cause, in particular to determine how far they can go to defend perceived rights or avenge perceived wrongs. There are many rules that are more or less binding, many customs and mores with real force, many widely accepted concepts of right and wrong, but there is no solid consensus to which all men of good will (much less men of evil intent) will submit, and which can therefore be used to adjudicate or enforce various interpretations of what is and is not acceptable behavior.

While the practices of both primitive tribes and of the current international system can be described as conducive to the outbreak of violence, in practice both systems create more periods of peace than of conflict, and more peaceful settlements and de facto compromises for confrontations than eruptions to all-out conflict. While there may be various councils and international organizations, there is no genuine legislature and no central authority with the right, obligation, and power to enforce even the most widely accepted values, customs, mores, and rules. Instead they are imposed by a kind of general will of the community, by self-help on their part of the injured party (or by allies or others on whom certain special obligations fall in various situations), and by a great deal of deterrence and restraint—much of it self-deterrence and self-restraint. With all of their defects, risks, and uncertainties, both systems work fairly well. This implies a certain, but not necessarily overwhelming, value in preserving much of the current international system, and even the present global political status quo.

Thus, it is because the structure of power and the techniques for preserving law and order in traditional societies and the modern international system have many similarities that we have turned to the collective wisdom embodied in much of human history for inspiration and examples. This is particularly true with regard to the concept of *lex talionis*, which we take up in the next section.

The purpose of the examples and analogies used here is less to improve the reader's knowledge of history or other cultures than to provide a vivid and interesting way of illustrating and explaining major issues related to nuclear war. So long as the illustrations are realistic, make correct points (or sometimes just raise interesting questions), and are effective pedagogically, controversies about the historical or sociological record and its interpretation are secondary considerations. The main function of the exercise is to increase our understanding of contemporary military and political reality; it would carry us much too far afield to pursue the subject more deeply from a historical or other scholarly perspective.

One means of acquiring some synthetic experience about nuclear war is through *Gedanken* ("thought") experiments. Physicists and philosophers often design and analyze hypothetical experiments that they do not intend to do, and perhaps no one will ever do, but that, nevertheless, could conceivably be carried out. If one tries to apply this technique to social or political issues, there is a danger that the formulation may determine the lesson taught; and since the formulation may be misleading, even though plausible, the answer could be equally misleading.

In the following *Gedanken* experiments there is also a problem of bias. Most of them were conducted with audiences I happened to brief in various NATO countries. Thus, the results do not reflect a scientific sampling. Within each country the audiences varied widely, but except for those in the United States, they were normally comprised of military and civilian officials or defense professionals. Rather surprisingly, these different groups often reacted in much the same way—at least eventually (i.e., they went through similar phases over time). This is reassuring, but by no means conclusive regarding the predictive value lecture-hall reactions might have for real-world situations. As far as the reader is concerned, these are *Gedanken* experiments conducted in a quite unscientific manner.

Nevertheless, these *Gedanken* experiments raise with exaggerated starkness some crucial issues and force the listener or reader to consider the consequences of alternative decisions and actions. It is essential to try and make nuclear war issues as "real" as possible, and to compel the reader to work through the problems himself, thus making him more aware of the complexities, ambiguities, and terrifying choices that may be faced by decision makers during a nuclear crisis.

A First Gedanken Experiment: Lex Talionis

One *Gedanken* experiment that I have used many times and in many variations in talks given over the past twenty-five or thirty years begins with the statement: "Let us assume that the president of the United States has just been informed that a multimegaton bomb has been dropped on New York City. What do you think he would do?" When this was first asked in the mid-1950s, the usual answer was, "Press every button for launching nuclear forces and go home." At that time (and into the very early 1960s), the concept of a firm balance of terror was (correctly) not fully accepted because the U.S. had great strategic superiority. An all-out attack, therefore, might have been a perfectly reasonable response. If the Soviets had not been fully alerted (i.e., had less than a week or two of strategic warning), the U.S. Strategic Air Command (SAC) could almost certainly have destroyed most of the Soviet strategic force on the ground, and the Soviets would have been able to launch very few bombers, if any, against the United States. (They might, of course, have been able to do very severe damage to Western Europe and they might have been thoroughly alerted—although it was not until the mid- or late 1960s that the Soviets achieved a serious alert status that would have enabled them to strike back or preempt effectively against the U.S. heartland without having—and giving—a great deal of warning.) Therefore, even if simply launching SAC would have been the wrong thing to do, it was not a completely ridiculous answer from some perspectives.

In any case, the dialogue between the audience and myself continued more or less as follows, after they suggested "press every button":

Kahn: "What happens next?"

Audience: "The Soviets do the same!"

Kahn: "And then what happens?"

Audience: "Nothing. Both sides have been destroyed."

Kahn: "Why then did the American president do this?"

A general rethinking of the issue would follow, and the audience would conclude that perhaps the president should not launch an immediate, all-out retaliatory attack.

However, in the mid- and late 1960s, before audiences in Western

Europe and Canada, I often still got much the same answer to the first question ("Press every button") leading inevitably to, "Both sides have been destroyed." On the very last question ("Why did the American president do this?") these groups would tend to say, "That's the American president for you!" There was no suggestion that their original answer was ridiculous and needed rethinking. Needless to say, I took this attitude as reflecting a mild form of anti-Americanism.

In the mid-1960s, almost all audiences in the United States came up with a very different initial answer, which was, "Get on the 'hot line' with the Soviets and find out if the bomb was one of theirs, and, if so, whether the attack was accidental or not." This was a perfectly reasonable answer (and one that we will pursue further).

By the early 1970s, American audiences gave a very different answer to the first question. They asked, "Is it one of our own?" This indicated rather low morale, but it is an attitude which, in the imagined circumstances, reflected a perfectly realistic concern.

In fact, the first thing the president would do is almost certain to be different from all of these answers. Before doing anything else, he would ask, "Is it really true? I don't believe it." He would insist on receiving a great deal more information and confirmation of the nuclear detonation before taking any further action—although he might place SAC on a higher alert (which would have happened almost automatically as a result of the signal that a bomb had exploded).

Let us go back to the *Gedanken* experiment and modify it. It is perfectly clear that a bomb might well drop on the United States, even on New York City. Presumably the Soviet do have missiles that are on alert and aimed at New York, and presumably one might be launched purposely or by accident. Nor is it inconceivable that U.S. military forces might somehow drop a bomb on New York City as a result of some accident, mistake, or bizarre, unauthorized behavior. But at this point, assume that the president does get the Soviet leader on the hot line and asks what happened, only to get an answer that he simply would not get in reality:

> It's one of ours. We did it deliberately. A large number of people have been pushing us around recently: the Chinese question our preeminence in the Socialist camp; you wage an ideological crusade against us; and our East European allies are getting rebellious again. Generally speaking, we don't feel the Soviet Union commands enough respect and we decided to correct that situation.

After this act of dropping a nuclear bomb on New York City, everybody will respect us, not in the sense of love and confidence, but in the sense of fear and terror. When we speak, people will listen; when we make requests, people will comply, and we think that that is a good situation. Therefore we dropped the bomb on New York City. It solves a lot of short-run problems. It may raise new ones, but that's life. In any case, as far as we are concerned, the matter is closed.

In the 1960s, almost nobody in the audiences I addressed believed that the matter was closed, and still thought the president would "push all the buttons" in retaliation. In the 1970s, some people did believe that the issue was settled. They argued that to retaliate in a small way would be not only ineffectual but counterproductive, and they did not believe that the president would retaliate on a large scale for fear of touching off an escalation to all-out central war.

But assume that the U.S. did retaliate in a small way, by exploding a low-yield nuclear warhead 100,000 or 200,000 feet over Moscow (creating a fantastic fireworks display, but doing little or no damage), by launching a missile at a remote area in the Soviet Union (to destroy, say, a gaseous diffusion plant), by insisting that the Soviets pay some kind of indemnity (with the Soviets indicating they were willing to discuss the matter at a conference table), by breaking off cultural or diplomatic relations, or by persuading the UN to pass a resolution condemning the Soviets.

In this branch of the *Gedanken* experiment, any of these or other (equally ineffectual) retaliatory measures could be taken. What would happen next? The Soviet judgment that prompted the attack would be vindicated: they would be terribly feared and we would look weak, even if we launched a major mobilization program in response to the Soviet provocation.* The leaders of almost every industrial nation and most of those countries allied with us would think that "if the U.S. cannot protect New York City from capricious destruction by the Soviets, then we must make some kind of arrangement with the Kremlin, or prepare to defend ourselves." Every country in the world would have to either develop its own nuclear deterrent or come to some kind of modus vivendi with the Soviet Union. Many would assume a foreign-policy stance corresponding to "Finlandization," or become even more submissive to Soviet demands.

---

* Issues regarding a U.S. strategic mobilization are examined in chapter 7.

These far-reaching ramifications of the Soviet attack suggest that in addition to considerations of justice, honor, revenge, indignation, and anger, there would be important prudential reasons to retaliate strongly against the Soviet Union. Indeed, it is probably a lot safer and more prudent to retaliate effectively than not to do so at all, or to retaliate in a small way.

Therefore, let us assume that we do not strike back in a small way, but are determined to retaliate effectively. What will be our response? In posing this question to an audience, I insist that at this point they go through their own reasoning. Listening passively to the sequence of events in the experiment is not the same as actively struggling with a bizarre, even if hypothetical, situation, in order to gain a greater understanding of more realistic problems. The people in the audience must really imagine that they are the president and try to face the choices he would face. If they do this, they are really able to think through some of the issues.

When prodded (but not coached), most of the audience arrives at more or less the same solutions: either a massive attack on Soviet military forces or cities, or a very limited tit-for-tat strike. The second reaction is *lex talionis,* which is sometimes known as the law of "an eye for an eye and a tooth for a tooth." *

A slight digression on this talionic retaliation is in order. Every primitive tribe, in the absence of a legislature and courts, seems to have adopted some form of *lex talionis* as part of its legal system. The Book of Exodus in the Old Testament gives a very clear injunction calling for *at least* an eye for an eye and a tooth for a tooth (in order to preserve the law), but also *at most* an eye for an eye and a tooth for a tooth (in order to restrain the nature and extent of the counterattack). Even if severely provoked, one is not entitled to use excessive violence in his retaliation, thus becoming an aggressor himself. The punishment must fit the crime.

Lex talionis is a peacekeeping strategy very similar to that employed by the United Nations. When the UN intervenes in a violent situation, it usually does not try to determine who is right or even to propose a genuine solution for the underlying problems. Instead, it seeks an immediate halt to the violence, and only afterward tries to resolve the issues behind the conflict.

---

* It is interesting to note that much fiction as well as scholarly analysis has used proportionate (tit-for-tat) response as being the only plausible response in situations in which the only other apparent alternatives are holocaust or an effectively passive acceptance of an extreme provocation.

Members of the primitive tribe notice that it is very difficult to stop violence as long as there is inequity in the amount of damage suffered by one side or the other. Western societies generally regard any kind of tit-for-tat system for settling disputes as violence-breeding because both sides must more or less agree on whether the tit-for-tat rule has been satisfied. A primitive tribe, which has very few methods of enforcing the peace, finds that tit for tat works a good deal more often than it does not. They therefore accept it, not as an ideal method of keeping the peace (it certainly is inferior to the governance by a legislature and an executive in more organized societies), but as a great deal better than nothing.

But let us return to the *Gedanken* experiment. Some members of the audience, particularly if they have not thought much about these problems, recommend striking back at the city of Moscow. Those who are more knowledgeable, including most so-called hawks and SAC generals, almost invariably pick Leningrad as the proper target. Their reasons for selecting Leningrad are very clear: Moscow is more important to the Soviet Union than both New York and Washington are to the United States. Moscow is the central city of the Soviet Union. Leningrad is one of the great cities of the world. It has the Hermitage Museum, great schools, and distinguished writers, artists, scholars, and scientists; it is probably the most cultured and attractive city in the Soviet Union. It also has a population of four-and-a-half million, almost none of whom were directly involved in the Soviet decision to attack New York. Nevertheless, there is general agreement that in these bizarre (and admittedly unrealistic) circumstances, the destruction of Leningrad would be the best possible reaction—perhaps in a manner that would preserve the lives of many of the inhabitants (e.g., by permitting the preattack evacuation of the city). The U.S. government might conceivably feel that the people of Leningrad were not implicated in the decision by their government to destroy New York City. Therefore, American leaders might want to punish the Soviet regime by destroying an important national treasure, while doing as little damage as possible to more or less innocent bystanders.

Most people would argue, however, that citizens must suffer the consequences of acts taken by their government, whether they were explicitly consulted or not. One almost has to adopt such an attitude, even if it is close to the edge of immorality. There is almost no other rule that would work—at least in most such circumstances. Of course, in this particular (imagined) scenario, the United States, by sparing most of the populace of Leningrad, would emphasize the difference in values between the U.S. leadership and the Soviets' ruling elite. Whether this approach is good or bad in its consequences is very controversial.

Assume that Leningrad is leveled by a U.S. retaliatory attack. What would happen after that? Almost certainly nothing—at least in the short run; there would be an acceptance of the fait accompli. The Soviets expected Leningrad to be destroyed when they attacked New York City, though naturally they did not say so. Their acceptance of the U.S. retaliatory attack would be independent of whether Leningrad's inhabitants had been killed (unless, perhaps, the Soviets had successfully attempted to spare New Yorkers).

We now pursue a slight variation of this scenario, namely, that the United States decides to retaliate against Moscow. Now what seems likely to happen? The surviving Soviet leaders might then get on the hot line and say, "Don't you read your own books, or believe your own theories? We understood that if we bombed New York, you would retaliate against Leningrad. Well, you have now escalated, and escalated very severely. We have to teach you not to do this." In principle, the Soviets might launch an all-out attack, but if they did, the U.S. would retaliate. They might want to obliterate Washington to equalize the damage, but this would probably seem to them too likely to prompt at least commensurate retaliation and perhaps further escalation. Yet they could hardly ignore such a flagrant "violation of the rules" by the United States.* One can easily imagine them instead choosing to destroy a lesser city than Washington, one which was important enough to be significant, even if it did not quite even up the damage. New Orleans or San Francisco might be the target of the next Soviet strike.

What if New Orleans or San Francisco were destroyed? What would we do then? Usually the audience at this point says, very reluctantly, "Nothing." They can easily imagine this "city-trading" conflict going on indefinitely if the U.S. retaliates again, or even erupting into all-out war. I believe this view is basically correct. The Soviets committed the original provocation—destroying New York City—for whatever reasons they may have had for taking such a bizarre action. By destroying Moscow, the U.S. taught the Soviets that they could not get away with this lightly, and thereby made it unlikely that they (or others) would quickly resort to this tactic in the future. When the Soviets destroyed New Orleans or San Francisco, they made the point that they had as-

---

* These rules are not embodied in any treaty—or even in any governmental pronouncements—but they exist, even if only in an ad hoc fashion, though one that clearly applies in this hypothetical confrontation.

sumed they would have to suffer some retaliation, and had already taken this into account, calculated the cost (including the risk of escalation to an all-out war) and correctly or incorrectly found the whole enterprise to be worthwhile. But by hitting Moscow, the U.S. had increased the level of violence far beyond that of the initial provocation, and far beyond what the Soviets thought they deserved. They now had to teach the United States (and the world) not to do such things. They may also have had to satisfy their own people that the United States could not get away with a disproportionate counterstrike. While the Soviets had not quite evened up the consequences of this competitive escalation, they had done enough to make all the necessary points.

Finally, by failing to retaliate against the second Soviet attack, the United States would not show any great weakness. It is very clear that the Soviets would have suffered more total damage than the United States. Thus, most of the world, including the United States, would see the logic behind the Soviet actions. Though the U.S. would have been "justified" if, in response to the destruction of New York, it had escalated to an all-out war, by choosing to hit only Moscow it had somehow made clear that it was both "playing the game" by the rules and violating those rules at the same time. Thus it would seem that in this *Gedanken* experiment there was a kind of logic to the situation based on *lex talionis.* Anything other than the U.S. attack on Moscow, followed by the Soviet retaliation, would have either left the matter unsettled or have had an all-too-good chance of exploding into total war.

In this example, we have assumed that there was a firm balance of terror, or a condition of mutual assured destruction. As a result, the U.S. could not launch a counterforce second strike that would be sufficient to prevent the surviving Soviet strategic forces from being used to destroy many additional U.S. cities. Instead, the United States had to constrain its response to what is sometimes called a "limited nuclear option" (LNO). To the extent that a balance of terror did not hold, then a very basic assumption would be relaxed and an LNO might not be appropriate in the postulated circumstances; it would be too dangerous if the imbalance favored the Soviets, and the Soviets would probably have used a lesser level of violence to make their point if the U.S. had strategic superiority.

This is one reason why the U.S. might eventually want to acquire what I refer to as a "not incredible counterforce first-strike capability." The notion here is that if the Soviets sufficiently provoke the U.S., we may prefer trying to disarm or otherwise weaken them militarily through some kind of limited attack directed against their strategic nuclear forces. Such an attack would avoid enemy urban-industrial concentrations in an attempt to use cities and citizens as hostages to prevent or limit retaliatory attacks against the U.S. population. Unfortunately, at the present time it seems likely that the Soviet Union, not the U.S., will have the ability to threaten a not incredible counterforce first strike in the early and mid-1980s, although the United States will acquire a comparable capability in the late 1980s and early 1990s (when the current strategic force modernization program is completed).*

The bottom line is that if a nuclear-armed nation facing a nuclear-armed opponent does not have the ability to initiate and survive a nuclear war when it is severely provoked, then it may have to consider some strange alternatives. A talionic retaliation (e.g., an LNO) is one substitute for an all-out war. In many crises it may turn out to be the "least undesirable alternative," † certainly if the confrontation were kept to a relatively low level of violence. The Soviets would be much less likely to provoke us at almost any of the levels considered in this chapter. And, of course, with good reason.

A not incredible counterforce first-strike ability would, in most cases, make it impossible for a member of the Soviet Politburo or an officer of the General Staff to argue persuasively that the U.S. really would restrict its retaliation to the destruction of Leningrad; or that if the U.S. exceeded the limits of its strategic theory and the unwritten "rules of the game" and destroyed Moscow instead, it would then restrain itself following a second Soviet attack. In the long run, it may be very important to the United States to appear, with more than a modest degree of credibility, to have such a first-strike capability, even if that capability could not really be considered very credible (for on these issues even a little credibility goes a long way).

Let us now turn to an even more stark *Gedanken* experiment, but one which, in my judgment, is equally sobering and informative. This intellectual exercise makes even clearer the need to have either a not incredible counterforce first-strike capability or an understanding of talionic retaliation (or other high-level tit-for-tat possibilities).

---

* See chapter 6.

† The simpler and more natural phrase "the most desirable alternative" undoubtedly would offend many readers by not emphasizing that no one really likes the idea of using an LNO, only that they may not have any other options. This can be true in practice even if a government has rejected the idea in peacetime (as the Soviet regime apparently has).

A Second Gedanken Experiment:
Deterrence Can Be a Two-Way Street

I have used this particular "experiment" with a number of different audiences. One was with a group at an off-the-record meeting that took place at the University of Chicago Public Affairs Conference Center in 1963. (I believe that I am not violating any serious confidences by relating some of the dialogue twenty years after the fact.)

The conference was attended by a number of prominent Americans: business leaders such as Thomas Watson (chairman of IBM), philosophers like Professor Leo Strauss, government officials such as Adam Yarmolinsky (a special assistant to Secretary of Defense McNamara), four members of Congress, several generals, and many civilian defense analysts. The purpose of the conference was to discuss East-West tensions and the related danger of nuclear war.*

Many conference participants felt at that time that the Soviets would clearly be deterred from provoking the U.S. if we had nuclear retaliatory forces capable of inflicting "unacceptable damage" on their country. However, the conferees seemed to have rarely, if ever, asked themselves some very relevant questions: "Would U.S. leaders have the will to launch such an attack? What if the Soviets could inflict 'unacceptable damage' in retaliation? Would we not then be deterred from utilizing these powerful forces of ours, even if we were seriously provoked?" In this regard, French President Charles de Gaulle had already made the possibly prophetic comment (which I paraphrase), "It is easy to imagine a melancholy day when, in a war, the Americans destroy Eastern Europe and the Russians destroy Western Europe, but neither belligerent touches the other's homeland for fear of the ensuing nuclear retaliation."

I decided to elicit the audience's reactions to a stark hypothetical situation, in order to get quick agreement on certain elementary concepts, in particular the notion that de Gaulle's concern, while excessive at the time, was certainly legitimate—especially since he had the long-term future in mind. I presented the following deliberately shocking and extreme scenario: The Soviets destroy five capital cities in Western Europe—London, Paris, Rome, Stockholm, and Bonn, just to show that they can get away with it. The U.S. president then calls a meeting of

---

* Revised versions of the papers presented at the conference were published in Robert A. Goldwin, ed., *Beyond the Cold War* (Chicago: Rand McNally & Co., 1965).

the National Security Council and says the following: "We have to draw a line somewhere and the Soviets have clearly crossed that line. The time has come to 'push the red button.' "

The president is then informed by his defense advisers that if he launches an all-out nuclear strike against the Soviet Union, the attack will kill every single Soviet civilian. However, U.S. strategic forces will be unable to eliminate Soviet strategic forces. One cannot even pretend that a counterforce attack will be successful because the Soviets have hidden their forces under the ocean, made them airborne, or encased them in protective silos. Therefore, they cannot even be located effectively and targeted. Soviet nuclear forces are unquestionably invulnerable to a U.S. strike. There is absolutely no military means of preventing the Soviets from retaliating. (Of course, it is quite difficult for the Soviets to hide and protect their strategic forces so effectively in the real world. Indeed, I will argue that it is very important for the U.S. to design its strategic-force posture so that the assumption in the *Gedanken* experiment does not hold.)

The president is also told that the Soviets will not hesitate to strike back in response to any devastating U.S. attack, and that their retaliatory blow will kill virtually every American citizen. Finally, his advisers report that his only attack option is to unleash *all* of the U.S. strategic force. U.S. political decision makers and military planners had not bothered in peacetime to think through any other strategic war plans (e.g., LNOs). Moreover, the Defense Department has not procured forces and command-and-control systems with flexibility sufficient to permit last-minute changes in strike plans. (Again, while these last assumptions are part of the *Gedanken* experiment, they are not completely unrealistic with regard to the real world of yesterday and, to some extent, even today.)

I then asked the participants the following questions: "Would the president order an attack by U.S. strategic forces? Should he do so? What would you do?"

Only four people present said that the president would authorize an attack. These four also thought that he would be right to do so, and that they would do the same if they were in charge. No one else took any of these positions. The four individuals were the four congressmen: Senator Bourke Hickenlooper, Senator Henry Jackson, Representative Melvin Laird, and Representative Gerald Ford. Later, of course, Secretary of Defense Laird and President Ford held very responsible offices with respect to the issues raised here, but undoubtedly changed their minds when they assumed those duties.

Although many of those in the audience thought of themselves as extremely conservative and tough, even hawkish, on national security matters, no one else at the conference agreed with the four. Senator Jackson (with whom I had been quite friendly) said, "Herman, I thought you were a hard-liner!" I replied, "Scoop, I am, but not *that* hard." A very spirited discussion ensued. Everybody else present thought that the president would not order this militarily useless reprisal. They believed that, given the assumptions of the "experiment," if the president was also in good health and of sound mind, the Soviets could rely on an unwillingness on his part to authorize that particular attack. The president would doubtless approve some alternative response, but he would not make a decision that would, in his judgment, inevitably result in the total destruction of the United States—no matter what was called for by honor, treaty, or plain anger. The situation was not analogous to that of a soldier dying for his country, a platoon dying for an army, or an army accepting annihilation for some national interest. In a sense, there is nothing that an entire country can "die" for that has a larger value than the continued existence of the country itself.

One individual suggested that if somebody attacked his wife he would not hesitate fighting that man, even if the assailant was considerably stronger or armed with a knife. I asked him if his reaction would be the same if the assailant brandished a machine gun that he was fully prepared to use. At this point he backed down. An enraged husband might strike back at someone who was humiliating his wife, even if the attacker had a machine gun, but the president of the United States is not likely to give way to emotion in a crisis prone to nuclear escalation. He is much more likely to become cool and more sober. Nuclear weapons, so far from being an inducement to indiscriminate slaughter, may be a very powerful inducement to rationality, even though they may be difficult to employ rationally. Thus, the antipathy to nuclear weapons, the very "unthinkability" of nuclear war, may be a reason to think that the antipathy to nuclear weapons will survive the first burst.*

The idea that the U.S. can be deterred even if gravely provoked is

---

* The dangers and possible consequences of a thermonuclear war may appear so stark to both U.S. and Soviet decision makers in an intense crisis, that national and personal idiosyncrasies would become swamped by nuclear fears, and the behavior of the two leadership groups would approximate that of what might be called the "bourgeois businessman model" of nuclear war. This model postulates that leaders will be so frightened of the initiation or continuation of nuclear conflict that their main objective will be to end the crisis or fighting on the basis of the first set of acceptable conditions that arises. No effort will be made to bargain for further advantages. Nor will there be any attempt by a defeated nation to fight on with nuclear strikes launched in desperation.

no longer as hard for most Americans to understand, but it was remarkably difficult for many to do so in the early 1960s. I would not need to resort to such a harsh scenario to get this idea across to a typical audience today. Nevertheless, on the whole, many people—including congressmen—still do not understand this basic issue or many of its nuances (or at least do not choose to understand it).

To give some credence to this assessment, I would cite the following exchange, which took place between myself and Senator William Proxmire during a congressional hearing held in 1975:

Mr. Kahn: . . . let me set a context for [a constrained Soviet counterforce] attack. The Soviet Union is striking the United States. They are very anxious for us not to strike back. Even if they have been evacuated, they don't want their cities destroyed. Those buildings in Moscow and Leningrad are important to them.

Now, imagine two Soviet planners. One says, "I want to maximize the military efficiency of the attack." But then, 18 million [Americans] get killed, or make it 50 million people. It improves my argument.

The second planner says, "We can take a 10 percent military disadvantage and kill essentially nobody by fallout. Absolutely nobody and, therefore, we can hold all those people as hostages to increase the probability of good behavior by the United States."

Which planner is going to win that argument?

There is no question in my mind it will be the second military planner. Anybody that doesn't understand that, simply shouldn't be commenting on this kind of material.

Senator Proxmire: That still eliminates a lot of people.

Mr. Kahn: I agree with that.

Senator Proxmire: Don't you think it is unlikely that the Soviets could possibly envision an atomic attack on the United States, that, as you say, would reduce by 10 percent their military effectiveness and in doing so, in order to minimize the effect on human life and maybe not kill 50 million, but kill a million or two million and not have an all-out devastating retaliation by this country? They know that.

Mr. Kahn: They don't know that.

Senator Proxmire: Well, I can't imagine a scenario. I can't imagine the President or Congress standing still for that, without immediate, terrific retaliation. Are we going to say we give in?

Mr. Kahn: No. There are other capabilities.

Senator Proxmire: What?

Mr. Kahn: The ability to destroy their military forces, increase bargaining or do something in return.

Senator Proxmire: You know that is going to escalate in a hurry. I don't think those are the only options, as you know. There are the retaliations that might be more limited to begin with, and the Soviet retaliation might be limited. These things are going to escalate fast. *The strategic value and military value of destroying cities in the Soviet Union would be very great.* We don't want to lose. They don't want to lose. [Emphases added.]

Mr. Kahn: I agree with the first part of your statement. There are other options. . . . We will do something to the Soviet Union, but we will not do anything that makes the destruction of 200 million Americans inevitable. No American President is likely to do that, no matter what the provocation.*

The dialogue illustrates a serious problem. Senator Proxmire is highly intelligent, well read, knowledgeable, and thoughtful. Nevertheless, in my view, he really did not have a firm grip on some fundamental concepts: (1) in initiating a nuclear war, the Soviet Union would not necessarily choose to maximize American casualties, since this would almost guarantee a corresponding U.S. second strike against Soviet civilians; (2) the U.S. might be deterred from massive retaliation against Soviet cities for fear of provoking a Soviet attack against U.S. populated areas; and (3) civilians and cities (as opposed to military forces and installations) would not be priority targets in many conceivable nuclear wars.

The point that deterrence can be a two-way street in both peace and war is so important, and sadly misunderstood, that it is probably useful to cite another anecdote that supports and elaborates the argument. Again the general story comes from briefings I gave in the early 1960s before groups composed primarily of senior American military and civilian officials. At the time, most of these officials believed that we could never be deterred from retaliating; thus the Soviets would be deterred from engaging in any extremely provocative acts (e.g., an invasion of Western Europe, a first strike against the U.S.) because our potential reprisal was so "credible," i.e., so probable and so awesome in its destructive power. To some degree they were right, given the balance of nuclear forces at the time. There was no question in my mind that American leaders would have reacted very forcibly, and that the Soviets understood this (in fact to some degree my audience was put off because

* Joint Committee on Defense Production, Hearings, *Civil Preparedness and Limited Nuclear War* (Washington: GPO, 1976), pp. 53–55.

my *Gedanken* experiments were based on clearly contrafactual assumptions regarding the U.S.-Soviet strategic balance). But the point was that we were predicting the reactions of U.S. decision makers over the next ten to fifteen years, when the balance of forces might (and in fact did) become quite different. (And even in the beginning of the 1960s, the topics I raised were becoming increasingly relevant.)

In briefings of this kind I would often start off by telling a well-known joke associated with George Bernard Shaw. The great playwright was supposed to have walked up to a beautiful woman and said, "Madam, would you sleep with me for a million pounds?" She hesitated, blushed, and said she would. He then asked, "How about for two pounds?" She slapped his face and said, "What do you think I am?" He replied, "Madam, I know what you are. We are simply haggling over the price."

I would then go through the *Gedanken* experiment in which the Soviets were assumed to possess an invulnerable strategic force and destroyed five European capitals just to show they could get away with it. Eventually, every high-ranking official in the room would admit that he would not retaliate, that the president would not—and should not—retaliate, and that the Soviets probably could rely on this.

Every now and then a young officer would stand up and say, "These opinions are both ridiculous and disgusting. I think we should retaliate. I would myself." I would then remark, "This young man is obviously of great value to the armed forces; he has the kind of high morale we want in a line officer or a fighter pilot. However, this particular briefing is for staff officers, for whom high morale is much less important than good analysis." (I would half-jokingly invite the young officer to leave the room.)

Next, I would ask the group if we would retaliate if we faced a risk of roughly 10 million casualties. In those days morale was still high, as well as a conviction that our international obligations must be honored. Almost everybody would say, "Yes, we would."

I would then say, "O.K.," and, referring to the Shaw joke I had told, add, "It is clear we are now simply haggling over the price." We would generally conclude that the United States would probably be deterred at the point where it faced between 10 and 60 million casualties from the Soviet counterretaliation (60 million being about a third of the population at that time), and that, of course, even if a U.S. nuclear response seemed merely "not improbable," it could nonetheless be very deterring to some Soviet decision makers (but less to others who might be willing to run some risks if under sufficient pressure in a grave international crisis).

One point in going through this exercise is to make plausible the notion that the problem of "acceptable level" calculations exists. The proper response probably involves some kind of limited strategic retaliation or talionic response. Suffice it to say that there are great risks and costs attached to all these responses, even if desperation suggests their use.

The lesson here is that the threat of retaliation, and thus deterrence, is relative and depends to a great extent on the price to be paid (as well as the nature of the initial provocation). The failure of the United States to develop more extensive and effective damage-limiting capabilities (e.g., ballistic missile defense) greatly increases the likelihood that American leaders will be deterred—or at least very restrained—in a range of possible nuclear confrontations with the Soviet Union.

A Third *Gedanken* Experiment: Ramifications of Soviet Superiority

A few years ago, I served as chairman of a panel on "strategic nuclear stability" at a meeting sponsored by the National Defense University.* Many panel members were thought of (by themselves as well as others) as being rather hawkish on defense issues. I decided to demonstrate that these so-called hawks were not paranoid about the Soviets and at the same time to explore some of the implications of potential Soviet strategic superiority in the 1980s.

I asked the panel to imagine a situation which is not at all implausible (unlike the previous two *Gedanken* experiments). It involved a series of assumptions. First, ideological ones such as: the Soviets believe they have a historic mission to further the transition to worldwide socialism; they also believe that this final transition may well be preceded by a last desperate attempt on the part of the capitalist world to preserve its existence by launching an attack on the Soviet Union; they feel they can win such a war, but despite their rhetoric, they are not certain; on the other hand, a *temporary* "window of opportunity" has opened up

* See the Report of Panel One, "Toward the Maintenance of Strategic Nuclear Stability," in *Toward Cooperation, Stability, and Balance* (Washington: National Defense University, 1977), pp. 9–21.

which might make possible a much earlier and safer transition to a world socialist state. The group was also asked to make the realistic technical assumptions that the Soviet Union could destroy virtually all of the U.S. land-based ICBM force (while using a relatively small number of its own weapons), disrupt or destroy a very large portion of the U.S. bomber force, and reduce the capabilities of U.S. strategic submarines by some uncertain amount.

I then asked the panelists whether, under these circumstances, they thought that Soviet leaders would accept the calculations of their defense planners and deliberately launch a first strike against a seemingly weaker United States. With one exception (a physicist, not a Soviet affairs expert), all the panelists thought that the Soviet Union would not deliberately plan to launch such an attack more or less out of the blue.

The point is not only that they would simply suffer too much damage to their own cities and population in return; they believe they could recover from this damage, and very likely they could. (This would be especially true if the Soviets were able to dominate the postwar world, or at least work under advantageous terms with most of the surviving nations.) The Soviets, in addition, would have an extremely difficult time justifying their action to those other nations and to their people, since there would have been no prior crisis or perceived confrontation with the United States. Also, under these circumstances, the Soviets could not evacuate their cities effectively (for fear of warning the U.S. of the attack) and therefore might well suffer 20–40 million immediate dead, even if the U.S. could mount only a limited, disorganized retaliatory strike.

Of course, the Soviets might try to protect their cities by "intrawar deterrence" (a tactic alluded to in the last section). That is, they might spare most of the large U.S. cities and then hold them as hostages to prevent an all-out attack on their own cities. While such a tactic is probably more technically and politically feasible than most commentators realize, it is by no means foolproof. If it failed, the Soviets would then face a very grave threat, though they could probably survive it. Still, one could easily imagine some kind of internal revolt or major change in the Soviet Union as a result of such a series of events that could clearly be blamed on the Soviet leadership—especially if any elements of the plan went very wrong.

However, the Soviets can do much better than this (or be forced into a position where they are inadvertently put into a better position). Imagine, as the NDU panel was asked to do, that the Soviets precipitate a crisis or take advantage of a spontaneous crisis because they have contingency plans for so doing. The crisis appears to both those in the know and outsiders to be of such a nature that neither side is really much more to blame than the other, but neither side feels it can afford to back down. However, both sides are very anxious for the other side to back down. To accomplish this, each side must create some kind of a theory as to why the other side should capitulate. The Soviets create exactly such a theory for the Americans and exploit the NATO allies' fear of war to put great pressure on the Americans to be reasonable.*

In this scenario, the Soviets announce to the Americans (and perhaps to their allies as well) that on "Saturday at noon," they are going to start evacuating their cities, and that this evacuation will be completed in about two days (resulting in a condition where 95 percent of the Soviet population is sheltered in areas outside of the urban centers).

If the Soviets succeeded in their evacuation and then struck against U.S. military targets, but in a careful and restrained fashion, they would probably kill fewer than 5 million Americans, perhaps many fewer. This type of attack would, to be sure, force them to accept some military handicaps (as I noted in my dialogue with Senator Proxmire). They would have to avoid some military targets, and some they would attack with less than top efficiency. But their forces held in reserve would then be substantially larger and more powerful than the surviving U.S. forces.

In addition, U.S. surviving forces would have been disrupted by the attack and forced to operate in a completely unfamiliar postattack physical environment that would present many problems, some of which would be partial or even total surprises. Because of this, the U.S. might, in fact, be almost completely incapable of retaliating. If, despite its difficulties, the U.S. did retaliate effectively with its remaining forces and struck at Soviet cities, it would obviously do an enormous amount of damage to these cities. But it is difficult to believe that the surviving forces would kill as many as 5 million Soviet citizens.

The Soviets would then presumably strike back at U.S. cities; the

---

* I know of no European ally that is actually prepared to fight a nuclear war. All of them would prefer accommodation to almost any degree of conflict. And they have reasons for feeling this way. Although the nations of Western Europe, like the United States, are targeted by hundreds if not thousands of Soviet nuclear warheads, the Europeans know that the Soviets are more likely to respect the territorial integrity of the United States (to prevent U.S. strategic retaliation) than that of Western Europe. This also applies to restraint in attacks on cities and civilians. Thus our allies may well be willing to back down even when we are not; in fact, they may force us to back down.

number of Americans killed would then depend on how successfully the U.S. had managed to improvise an evacuation and protective shelters —but it could hardly be less than 20 million and could easily be over 60 million. And the destruction of U.S. cities would be much more thorough than the destruction of Soviet cities.

Further, the Soviets would still have an enormous military advantage after the U.S. second strike. They would also be in a good political and moral position. They had given the U.S. an opportunity to back down from what would look like a clearly untenable position. They would have used their weapons in their first attack appropriately, carefully, and prudently against military targets, even at some cost in military effectiveness. The U.S., not the Soviet Union, would have broken the (implicit) rules by attacking civilians. Under these circumstances, the Soviets would have relatively clean hands in the eyes of their own citizens and the world, and a very good chance of being able to organize a worldwide recuperation effort. External assistance (provided as the result of trade or Soviet coercion) would almost certainly make it relatively easy for the Soviets to rebuild rapidly. Soviet postattack strategic superiority would make it almost impossible for the U.S. to recover if the Soviets wanted to keep the U.S. down.

Despite the possible advantages the Soviets could achieve by manipulating a crisis to the point where they could "justifiably" launch a first strike and disarm the United States, the panel still did not believe, with the same one exception, that the Soviets would deliberately make such plans. They all thought that the Soviets would regard such a move as "adventurism." Thus, even hawkish American analysts believe that Soviet leaders are essentially cautious. Despite the great temptation which is being created by the trend toward strategic superiority over the U.S., the Soviet Union is not really likely to initiate nuclear war deliberately for positive political gains.

Like the panelists, I do not believe that the Soviet Union would be likely to give serious consideration to a deliberate, calculated first strike against the United States except under the most extreme circumstances. Cultural, ideological, and analytical factors indicate that the Soviet leadership would act with prudence and caution. The Soviet Union lacks a tradition of successful "wars by calculation"; Soviet ideology stresses patience and warns against recklessness in the long-term struggle with capitalism; and technical uncertainties as to how military equipment would actually perform in a nuclear war would prompt the Soviets to think very hard before initiating a nuclear attack. However, this is not to say that Soviet strategic superiority would be irrelevant, or would give the Soviet Union no advantage in its relations with the United States, even if no such attack occurred or was explicitly threatened.

To illustrate one such advantage, let us add a different twist to the preplanned basic scenario described above. Again, a very serious but *unwanted* crisis has developed, this time in Eastern Europe, and the West is providing moral or more tangible support to those seeking freedom from Soviet domination. The crisis begins to get out of control; unrest is spreading throughout Eastern Europe. The Soviets want to halt the spread of the crisis, but for domestic and bloc political reasons, they feel they cannot afford to back down; if they did, they feel that they would then be even more likely to lose control. The Soviets calculate that if they evacuate their cities and demand that the U.S. and NATO withdraw their support, the U.S. and its allies will have virtually no choice but to back down. Thus, they evacuate their cities in a cautionary and relatively unprovocative fashion; as a result, the evacuation is quite incomplete, but it puts them in a position to accomplish a complete evacuation quite rapidly.

At that point, the U.S. would be in a very bad position to launch *or threaten* a preemptive strike, unless a great many changes had been made in the present U.S. force posture, including the implementation of a substantial civil defense program. The U.S. would be foolish to stand firm in the hope that the Soviets were bluffing—since the Soviet Union might well either strike or carry out a more complete evacuation. The U.S. would have to begin negotiating, possibly making dangerous concessions out of a desire to save itself or to buy time.

The difference introduced by this twist in the scenario is that the U.S. would no longer face a cool, calculating Soviet Union. Instead, the Soviet Union is caught in a very severe crisis involving a potential loss of bloc political control. There is a world of difference between acting to achieve political gain, and doing so to prevent intolerable political loss. If the Soviet Union possesses strategic superiority and significant defensive capabilities, Soviet leaders would be able to manipulate the crisis in a way that the United States could not—or even be tempted to solve the problem once and for all. If Soviet evacuation were achieved, even at the level discussed above, the U.S. deterrent threat of massive retaliation against Soviet cities would be greatly diminished —U.S. strategic forces would mainly hit empty cities. Further, the Soviets could turn the screw by the degree and quality of the evacuation and by the explicitness of their ultimatum and actions. (The role of evacuation in crisis bargaining is a subject taken up again in chapter 8.)

Thus, by being able to threaten credibly, or actually carry out an evacuation of its cities to one degree or another, the Soviet Union would be in a position to dominate any lower-level crises and confrontations that might occur.

Another possibility is even more ominous, if less likely. The Soviets might actually launch a symbolic or small-scale attack—especially if a confrontation had reached the point where Soviet cities had actually been evacuated—in order to demonstrate resolve to the U.S. and encourage U.S. accommodation. At that point, the Soviet civilian population would be relatively protected and U.S. urban-industrial assets would remain vulnerable. Further, if the U.S. retaliated tit for tat, the concept of intrawar deterrence might look to the Soviets as if it had now been tested and confirmed. Under such conditions, the world might look very different to the Soviet leadership than it had in the precrisis political and military environment. Despite all the uncertainties I have mentioned, the Soviets might well believe much of their own rhetoric, and actually feel that a nuclear strike against the U.S. would not produce insurmountable destruction or intolerable loss of political control. Thus, a Soviet first-strike threat might easily have a self-fulfilling effect: the Soviets could convince themselves that the final confrontation between socialism and capitalism had arrived, and that they would undoubtedly emerge victorious.

In any event, the key point of this third *Gedanken* experiment is that the Soviets are not likely to want to risk their gains of the last sixty-five years and the eventual (to their minds) triumph of socialism by initiating a deliberate attack against the U.S. simply because of a number of clear military advantages they might possess. That is, they would not launch a calculated first strike, *unless they had virtually no other alternative.* If nuclear war ever occurs, the odds are very high that it will come as a result of an unplanned, unwanted crisis in which the resort to nuclear war emerges as the least bad of many bad alternatives—i.e., when the risks or potential losses of not fighting are greater than the risks or potential losses that nuclear war would seem to entail.

Nonetheless, given the advantages the Soviet Union could reap from strategic superiority, I believe that it is dangerous (and somewhat absurd) that the United States, the wealthiest and technologically most advanced nation in the world, should not also be the strongest in most important components of military power; it should certainly not be forced to rely to such a degree on the prudence and caution of the Soviet leadership for its own security.

To put it in simple and logical terms, what the U.S. must be able to do in an intense crisis is persuade the Soviets that even though they prefer striking first to striking second, they should really prefer backing down to striking at all, because striking first is too likely to have extremely undesirable consequences when the U.S. retaliates (i.e., the Soviet regime might lose political control and be left with radically diminished military power, as well as witness severe destruction of its population and industry). This capability to coerce the Soviet Union during a serious crisis or a war is what I refer to as "escalation dominance." Escalation dominance requires that Soviet leaders always consider conciliation to be preferable to continued conflict and escalation. Such a criterion for measuring the adequacy of U.S. strategic forces calls for a much greater military capability than that associated with "assured destruction" (and flexible targeting options). However, as the *Gedanken* experiments in this chapter illustrate, our extended deterrence responsibilities with regard to the defense of Western Europe render escalation dominance a necessary standard for measuring "how much is enough" when it comes to the strategic forces this country procures and maintains.

PART TWO

Describing Strategic Realities

3

THE REVOLUTION IN WARFARE:
CONTINUITIES AND DISCONTINUITIES

This chapter considers some of the most important ways in which nuclear weapons have and have not changed the theory and practice of modern warfare and defense planning. We begin by noting that the idea that a "revolution" in warfare has taken place is imprecise; it gives the impression that the changes caused by the development of nuclear weapons occurred all at once. In fact, from 1945 to 1982, there were many "revolutions" in the technology and policies of nuclear warfare, bringing about changes as great as those that occurred between the Civil War and World War I, or between World War I and World War II.

The most fundamental of these "revolutions" clearly involves nuclear warheads and the increased efficiency in kilotons deliverable per pound of payload, going from .002 kt/lb for the Hiroshima bomb to a reported 2 kt/lb in the mid-1970s—a factor of 1,000 in thirty years, or ten in every decade, or more than three every five years. There have also been spectacular developments in ballistic missile technology, missile accuracy, nuclear submarines, antiballistic missile capabilities, cruise missiles, and so on. Piled one on top of the other, this progression has been so continuous it has appeared almost evolutionary.

The invention and development of nuclear weapons signaled the onset of the "nuclear era," but these revolutionary accomplishments would not have been nearly as dramatic had it not been for the many nonnuclear technological developments occurring at the same time. The debate over precisely what and how much has changed in the nuclear era, and how much of the prenuclear military wisdom is still valid, does not usually allow for the "mixed" position we take. Some students of military science argue that nothing much has changed from the days of the Greeks and the Romans—that war is still war, and its basic principles are the same; in particular, the objective is to defeat the enemy and compel him to surrender or otherwise accede to your demands. To accomplish this, a nation employs many of the same tactics and strategies that nations have always used. The opposing argument holds that traditional military concerns with the detailed strategies and tactics of conflict are irrelevant in the nuclear age; they claim that the only valid strategy today is total deterrence of war.

An extreme variation of the first position compares strategic forces today with the six-gun fighter epitomized in an earlier culture of the American West—whoever draws first and fires accurately still wins. An extreme variation of the other view is illustrated by the metaphor of two scorpions in a bottle—the inevitable result of a struggle is mutual suicide. Such phrases as "nobody wins a suicide pact," and "mutual assured destruction" are the typical rhetoric of this last position.

We believe that both of the above positions are largely incorrect or out of perspective. There are significant continuities with prenuclear warfare that military planners must still consider. But there are also significant differences that will influence developments in military procurement, doctrine, strategy, and tactics. In particular, we believe there is perhaps even more to be learned about the potential use of nuclear force in terms of the historical role of war than about any new uses of conflict. More than ever, there are lessons in the application of the nuclear threat "as a continuation of politics/policy by other means" (Clausewitz) and as an instrument for advancing the national interest by deploying military forces, though some important caveats and modification are needed.

If this discussion appears biased in favor of the continuities (rather than the differences) between the nuclear and prenuclear era, it is because the Western world has generally adopted such an intense belief in the discontinuities. Most people simply accept as obvious the idea that a nuclear war would be the end of history, and that it is therefore unproductive (or even counterproductive) to think creatively about strategy and tactics. From this perspective, thermonuclear weapons are the equivalent of the doomsday machine, an attitude that (while exaggerated) highlights one of the most fundamental changes between the prenuclear and nuclear eras—the unprecedented destructive potential of nuclear warfare.

A nuclear war, even if limited in scope, could easily cause between 10 and 50 million immediate fatalities, plus millions more wounded or ill. Such casualty levels have been reached before, but over a long period of time and among many nations (the estimated number of military and civilian fatalities during War War II was approximately 50 million). Coping with such huge numbers of casualties all at once would place tremendous (and unprecedented) strains on the resources and very fabric of society. Moreover, since these figures are estimated only from the known effects of nuclear explosions, it is possible that unknown and unanticipated effects could make the damage even greater.

Uncertainty is another significant new development of the nuclear age—the creation of weapons whose unknown effects may be more important, and more harmful, than the known ones. Except for Hiroshima, Nagasaki, and a limited number of nuclear tests, we have no recent or actual experience from which to make judgments or assessments regarding the use of these weapons, especially on the scale (thousands of megatons) that would be involved in a major war.

Because of the concentrated destructive power and instantaneous impact of nuclear weapons, the question of how a nuclear war begins is an important factor in determining its ultimate destructiveness. The initiation of the conflict is also a critical determinant of whether the attacked society can survive and ultimately recover. This, too, is a change from past eras when the outbreak of hostilities had little bearing on the war itself. If a nuclear attack were to come out of the blue, without any warning or opportunity for evacuation or other civil defense measures, the casualty figures could be far greater than if a nation's citizens were given time and opportunity to protect themselves. If a war were preceded by a period of heightened tensions or even conventional conflict, and if government leaders either officially ordered an evacuation of major urban centers or people spontaneously decided to take their own precautions, casualties could be reduced significantly.

The fear of nuclear weapons has prompted many dramatic slogans and rallies (as well as many fallacies and myths). In 1955, fifty-two Nobel Prize winners signed their names to the so-called "Mainau Declaration," which stated that "All nations must come to the decision to renounce force or war as a final resort of policy. If they are not prepared to do so they will cease to exist." Such positions summed up the typical

Western view that nuclear warfare was "unthinkable," and that the very survival of mankind was at stake as never before. A not uncommon thought, even among military generals, was that "if these buttons are ever pressed, they have completely failed in their purpose. The equipment is useful only if it is not used."

A letter written to me some years ago by Denis Flannigan, publisher of *Scientific American*, included the following:

> I do not think that there is much point in thinking about the unthinkable; surely it is more profitable to think about the thinkable. In my view . . . nuclear war is unthinkable.

All efforts, in other words, must be on prevention; none on fighting a war should prevention fail. Not surprisingly, advocates of this view believe that military preparations are stupid and dangerous. They scoff at attempts to reduce wartime casualties from, say, 100 million to 50 million, believing that any casualty level in that range is "unacceptable" and that the only acceptable option is the total elimination of war. They may disagree over which policies can best achieve this, but they share a horror and distrust for scenarios that suggest there could be a nuclear war in which one or both sides emerged with damage no greater than that suffered in wars of the prenuclear age.

Another position, which is not completely dissimilar but is much more acceptable to us, contends that in a reasonable world, the only rational and moral role for nuclear weapons is to deter, balance, and/or correct for the possession and/or use of nuclear weapons by an opponent. Unfortunately, the conclusion drawn from this defendable (if controversial) position often "degenerates" into a much less defensible assertion: that the only way to deter the use of nuclear weapons is through some form of mutual assured destruction. This deterrence-only position is actually quite close to that of the pacifists who also believe that war can be rationally eliminated—the former believe the constraint is fear, the latter believe it is a profound abhorrence of violence in any form. Both hope that the world eventually will disarm completely, but neither takes that possibility very seriously.

It is generally acknowledged that in the unlikely event nuclear weapons did become generally unavailable, a nation that somehow retained even a single weapon would represent a terrible threat to the rest of the world. Consequently, some supporters of disarmament simultaneously support a world government that would have a monopoly on nuclear weapons. The practical problems of this alternative, namely, the possibility that such a government itself could become oppressive, or could be taken over by an oppressive group, are rarely considered.

Another argument frequently made by nuclear pacifists and/or supporters of deterrence only is that war has become an inappropriate way to resolve international disputes in the nuclear age and an age of economic and technological interdependence. Modern developments have either obviated, lessened, or made transistory the strategic value of many geographic areas; nations can rarely gain wealth, commercial advantage, or political security by attacking a neighbor and conquering its territory. (The Middle East, and to some degree the Soviet bloc nations, may be exceptions to the rule.) Power, prestige, and influence are today attained primarily through successful economic development, not through military might or aggressive expansion.

While all of the above observations are true, the pacifist conclusion that "war never pays" is simply not correct. Wars have paid and paid handsomely in the past and could again in the future. In addition, the mere belief in an ultimate world without war does little to help resolve the tensions and strains of the current international system. To arrive at any kind of utopian accommodation in the future, we still have to get through the rocky and unstable 1980s and 1990s and had best be prepared to do so.

Deterrence-only positions can be dangerous. The most obvious question is—what happens if deterrence fails? Equally troublesome is what happens if there is a significant imbalance in the reliance on deterrence—if one potential adversary believes in it much more strongly than the other? One of the biggest changes of the nuclear era is that such fundamental philosophical differences about the nature of war can exist —differences about its prevention, its consequences, and its likely outcome. Based on the available evidence, it seems clear that the Soviet Union does not accept the West's apocalyptic view of nuclear war, nor do they support deterrence-only policies. Soviet military writings depict nuclear war as a survivable experience, and back up their reliance on deterrence with war-fighting capabilities, that is, the ability to fight a nuclear war and defeat the enemy.

Yet a belief in deterrence—including reliance on some degree of unilateral restraint in the acquisition, deployment, and role of nuclear weapons—has been a great deal more successful than most "realists" thought it could be (realists tend to discount the importance of voluntary unilateral restraint). Around 1960, I was predicting that by 1975 there

would be between five and twenty nuclear powers—twenty if nuclear
weapons had been used successfully on a limited basis (at that time U.S.
doctrine was to use tactical nuclear weapons under a very large range
of circumstances), and five if the (mostly "unilateral") psychological,
moral, and political restraints against such use remained in place.* In
fact, by 1975 there were only two additional known nuclear powers
(India and China) beyond the four that existed in 1960 (Britain, France,
U.S., U.S.S.R.). Nonproliferation policies have worked extraordinarily
well. It seems that they will work less well in the future. At least ten
nations have more or less openly indicated a significant interest in acquiring nuclear weapons in the next decade or two. Many believe that
Israel and South Africa already have such weapons; others think that
Pakistan and Argentina soon will. An explosive proliferation of nuclear
weapons sometime before the end of the century is not unlikely.

The voluntary self-restraint displayed thus far has been enhanced by
the cooperation of the nuclear powers that signed and enforced the
Nuclear Non-Proliferation Treaty, which along with the Test Ban
Treaty of 1963 is a prime example of relatively successful arms-control
negotiations. Arms control is by no means a new concept, but the great
importance placed on it—by both decision makers and the general public—is unique to the nuclear era. This is hardly surprising, given the
stakes involved. Everyone has an incentive to prevent the use of nuclear
weapons, and the superpowers are especially motivated to limit the size
(and the destructive potential) of their opponents' strategic nuclear
forces.

The environment for arms control is thus more favorable now than
it was in the prenuclear era, but unilateral military advantages (including
nuclear "superiority") can still be important, and incentives to cheat
still exist. Problems of verification remain, and are, if anything, exac-
erbated by the same advanced technologies that contribute to the sophistication of today's surveillance and monitoring capabilities.

In the past, some of the most effective arms-control agreements have
been unwritten understandings that effectively restrained opponents
from unlimited arms buildups. In the nuclear era this can still be the
case, since arms-control agreements need not necessarily be either
highly specific or carefully coordinated to be effective. In fact, such
preciseness is very difficult to achieve; the same fear and consequence
of cheating that make both parties extremely cautious and insistent that

---

* *On Thermonuclear War*, pp. 509–10.

negotiated agreements be spelled out in great detail can also make it almost impossible to achieve such agreements. Neither side wants to give the other an "institutionalized" advantage that could be decisive if deterrence should fail.

But deterrence has become the most basic politico-military strategy of the nuclear era. Bernard Brodie recognized that in the past the chief purpose of our military establishment was to win wars. "From now on its chief purpose must be to avert them." * To be meaningful, however, deterrence must take place in a specific context: who deters whom, from what actions, by what threats and counteractions, in what situations, in the face of what counterthreats and counter-counteractions? Some analysts argue that since we can do unacceptable damage to the Soviet Union, they will not provoke us; in fact, the Soviets may be able to deter our response to their provocation through threats or counteractions.

Deterrence, therefore, is not just a matter of military capabilities; it has a great deal to do with perceptions of credibility, i.e., the other side's estimates of one's determination, courage, and national objectives. For example, in the early days of the nuclear era, the British nuclear force probably could have inflicted much greater damage to the Soviet Union in either a first or second strike than the Soviet Union could have inflicted on the United States in a first or second strike. However, we are reasonably sure the Soviets were not too concerned about the British, whereas we were very concerned about the Soviets. The reasons, of course, have very little to do with theoretical military capabilities—they are a function of political realities. If the enemy is (correctly) convinced that you will not use your weapons, then it does not matter how sophisticated or powerful your weapons are. In many cases it is what allies and others think, or what they think the main opponents think, or what the opponents think the allies and others think, and so on, that is decisive.

This issue of perception and credibility often comes up in reference to the American commitment to NATO to deter a Soviet conventional attack on Europe through the threat of strategic retaliation. While the U.S. is clearly capable of such a response, it is increasingly doubtful that the United States would actually resort to nuclear weapons. The force of U.S. deterrence in Western Europe has therefore been substan-

---

* Bernard Brodie, ed., *The Absolute Weapon* (New York: Harcourt Brace, 1946), p. 76.

tially undermined by a visible weakening of U.S. resolve, a weakening clearly related to the buildup of Soviet intercontinental offensive forces, hence a revision in the strategic balance. Yet the Soviets still could not be "sure" that we would refrain from strategic retaliation, an uncertainty that would probably be sufficient to deter the Soviets from an attack on Western Europe.

The emphasis on deterrence in the nuclear age has led to a decline in the study and formulation of appropriate strategies and tactics for using the special qualities of nuclear weapons in guaranteeing the security of the U.S. When the atom bomb was first invented, many people felt that military strategy and tactics had virtually become obsolete since the inevitable result of a nuclear war would be world destruction, and that destruction would occur no matter what tactics were used. Tactical theory was therefore considered irrelevant. In addition, since the annihilation of the nation or the world could not be a national objective (strategy), strategy has become equally irrelevant. Clausewitz's concept of war as an instrument of foreign policy was no longer valid if war would destroy humanity.

The invention of the atomic bomb, therefore, seemed to end any constructive thinking about strategy and tactics. Nuclear war was simply unthinkable—both literally and figuratively.

This phenomenon, known as psychological denial, meant that while one side (ours) did little or no thinking about nuclear weapons, the other side simply regarded them as "bigger bombs," or "higher-quality weapons," and also did not undertake any fundamental rethinking of classical political and strategic assumptions. Indeed, on both the U.S. and Soviet sides, strategic concepts and tactics remained almost identical to those used in World War II. Attempts were made to correct the mistakes of World War II (e.g., the Strategic Bombing Survey prompted the selection of power stations as high-priority targets), but these "corrections" merely emphasized the absence of any new and creative thinking about the possible military value of nuclear weapons.

In the early 1950s strategic thinking had a partial revival. There was some discussion of the options open to a potential nuclear attacker— threats the attacker might make, tactics he might use if the threats failed, and the counteroptions available to the defender. For a time, the discussion reached a relatively high level of sophistication as it considered mixtures and levels of active and passive defense and of counterforce and countervalue targeting. But not for long. Strategic considerations came to an abrupt conclusion with the development of the H-bomb, which seemed so close to being a doomsday weapon that the details of war fighting really were irrelevant. Multimegaton weapons appeared to be unusable for any rational (and most irrational) purposes. There seemed to be no need for further strategic thinking.

In a well-known article, "Strategy Hits a Dead End," Bernard Brodie wrote:

> In a world still unprepared to relinquish the use of military power, we must learn to effect that use through methods that are something other than self-destroying. The task will be bafflingly difficult at best, but it can only begin with the clear recognition that most of the military ideas and axioms of the past are now or soon will be inapplicable. The old concepts of strategy . . . have come to a dead end.*

If anything, this position has become more extreme over time. A more recent commentator expressed much the same position:

> . . . the sheer destructiveness of nuclear war has invalidated any distinction between winning and losing. Thus, it has rendered meaningless the very idea of military strategy as the efficient deployment of force to achieve a State's objectives. . . .†

We would argue that both statements are incorrect. While both Brodie and Sigal admit that nuclear forces can and should be deployed for political and military purposes, they have difficulty seeing how a central nuclear war could achieve any policy objectives.

We believe the need remains for coherent and credible nuclear-use policies, including the need for clear and imaginative tactics. Terrible as nuclear weapons are, they exist and therefore may be used. Even if they are used only as a threat, such threats, if credible, in themselves represent a kind of use. When we deter the Soviets by the threat of escalation if they provoke a limited war, we are in fact using the potential application of our nuclear weapons. Even pure deterrence-only policies "use" nuclear weapons in the attempt to institutionalize a mutual paralysis through fear.

Military and civilian professionals understandably find it very difficult to deal with the impact of nuclear weapons on the actual waging of

---

* *Harper's* magazine, October 1955. (Nonetheless, three years later he published his opus, *Strategy in the Missile Age.*)

† Leon Sigal, "Rethinking the Unthinkable," *Foreign Policy* (Spring 1979), p. 39.

war and the handling of international crises. The uncertainty about nuclear war means that many alternatives and options have to be considered in the abstract. In addition, the complexity of nuclear issues and the awareness of the potential physical and political devastation caused by nuclear weapons have caused many strategic planners to opt out. But this is a mistake.

While nuclear weapons have certainly changed the ways in which wars will be initiated, fought, and won, they do not necessarily make war obsolete. As far as we can calculate, no plausible employment of current and likely future weapon systems would result in the end of the world or the end of the human race or anything close to it. But the fact that we have to add the caveat *"as far as we can calculate"*—and such phrases as *"plausible employment"* and *"likely future weapon systems"* is another example of how much has changed since these weapons were created. Nuclear weapons can do more damage than any other weapons in history, particularly if used in large quantities and in uncontrolled fashion.

But consider the possibility—both menacing and perversely comforting—that even if 300 million people were killed in a nuclear war, there would still be more than 4 billion people left alive. Studies of the likely casualty rates of nuclear conflict range from less than a million to some tens of millions (unless the war is totally uncontrolled, in which case the casualties might reach hundreds of millions). The worst-case scenarios of hundreds of millions dead and widespread destruction would be an unprecedented global calamity, but not necessarily the end of history. And a power that attains significant strategic superiority is likely to survive the war, perhaps even "win" it, by extending its hegemony—at least for a time—over much of the world. Indeed, throughout history there have been leaders who were willing to pay a great cost in national wealth and lives for a chance to take over the world.

While almost no one would take the position that thermonuclear weapons are "just another" advance in the technology of warfare, it is important to realize that these weapons, like others, may be fired. Accordingly, we would like to argue that one of the most important continuities of the nuclear era is that wars can still be fought, terminated, and survived. Some countries will win a nuclear conflict and others will lose, and it is even possible that some nuclear wars may ultimately have positive results (as World War II did). Reconstruction will begin, life will continue, and most survivors will not envy the dead. (Inhabitants of a country that loses a nuclear war and is very badly treated by the winner might envy the dead, but this, too, is nothing new in man's history.) It is important for people to continue to think of war as an experience that can be survived and recovered from if proper preparations are made, and important for national leaders to recognize that they will be judged by how well they exercise their responsibility to help their country prevail.

Like wars of the past, the extent and duration of nuclear wars could differ greatly, depending largely on the causes of the war and its political objectives. Nuclear war will not necessarily be an all-out strategic exchange, with each side firing most or all of its arsenal in a single spasm of destruction. It could well be limited, either by region and theater of battle or by constraints (either self-imposed or coerced) on the number of weapons used and/or the targets attacked. A conflict in the Middle East or Europe involving battlefield nuclear weapons would be a nuclear war, but would not be comparable to a nuclear war involving large-scale strategic strikes by the U.S. or Soviet Union against each other's territory. These two extremes, and the range of possibilities in between, require entirely different resources, planning, and preparations. None of them should be totally neglected if the U.S. expects to defend itself and its interests adequately.

Another similarity with wars of the past is that a nuclear conflict could occur over an extended period of time. History provides examples of extremely short wars (the Austro-Prussian War of 1866, the Arab-Israeli Six Day War of 1967) as well as extremely long ones (the Thirty Years War, the Hundred Years War and, though less prolonged, the two World Wars). The standard picture of nuclear war envisions a conflict of only a few hours or days, but it is possible that a war could last weeks or perhaps months. In fact, a number of military analysts are now arguing that a prolonged or protracted conflict with limited exchanges of nuclear weapons, followed by periods of bargaining and struggle for supremacy in limited theaters of operations, is a more realistic scenario. These analysts argue that neither side will use its entire strategic force in an initial attack, preferring to maintain a reserve force for postattack contingencies.

This new conception of nuclear war raises questions of command and control, as well as other aspects of endurance. It is not clear that strategic forces can be operated in a controlled and flexible manner in a poststrategic nuclear attack environment, nor is it clear that the forces can be maintained at high alert levels for long periods of time. However, the point we wish to emphasize is simply that, as with wars of the past, nuclear wars could last longer than generally imagined.

But issues of war fighting, termination, and recovery are much more complicated in the nuclear era than ever before. To a degree that really is unprecedented, the analysis of these issues depends on the examination of many different factors, as shown in Table 1.

In order to argue persuasively that it is possible to fight nuclear war, to survive, and to rebuild, one must be willing to argue that it is possible to handle every one of these issues in some adequate, or at least minimal, fashion. Many sober, competent, and distinguished observers claim this cannot be done, but what most of them probably mean is that they foresee terrible difficulties or problems in dealing with these issues. That is not the same as saying with certainty that survival is not an option.

An examination of the issues in Table 1 should make clear why so much attention is paid to the range of uncertainty regarding the outcome of a nuclear war. The unknowns can both increase and decrease the leverage of nuclear threats in peacetime negotiations. Stressing the dangers of escalation and the need for restraint by the opponent becomes a standard tactic. But is also reassures the weaker side that the stronger side is not likely to put great faith in the flawless execution of its plans; there is no track record of the weapons working reliably. Thus, it is often pointed out that war planning (a) must deal persuasively and convincingly with every point raised in Table 1; (b) will necessarily be both a difficult and complex process; and (c) is still likely to be highly speculative because of the uncertainties.

Table I
HOW MANY SURVIVORS?
WOULD THEY ENVY THE DEAD?

1. Prewar preparations (including lack of realistic "hands-on" experience)
2. Outbreak scenario
3. Immediate weapons effects such as
   A. Blast and prompt radiation
   B. Thermal radiation and fire
   C. Acute effects of fallout, electromagnetic pulse, earth shock, etc., and the likelihood of some unexpected weapons effects (including difficult operational problems)
4. War-fighting and war-termination scenario
5. Postwar scenario
   A. A reorganization period (including role of neutrals, Japan, and developing countries)
   B. Medium-term environmental problems

One typical ploy in a potential nuclear crisis is for one of the negotiators to point out the insanity of subjecting the entire world to the risks of nuclear war, regardless of the importance of the issues at stake to the two countries involved. Diplomats and military leaders have made antiwar arguments in the past, but the global dangers and risks are much greater in the nuclear age. The negotiator can urge a reasonable resolution by pointing out the disparity between some immediate crisis-prone "minor" issue and the almost infinite risks of a major war. Then he can suggest that "one of us has to be reasonable, and it isn't going to be me." This could end the bargaining and lead to war if his opponent's response is, "It's not going to be me, either." Or it could lead to capitulation by the opponent who realizes that compromise, or concession, is the only sane option, especially if the balance of forces is not on his side.

Some analysts, however, argue that superiority in the balance of forces can easily be dismissed as irrelevant in a nuclear war. This is probably too glib. In particular, the side with the advantage can argue that his government spent a great deal of money on these forces, something it would not have done if it thought they were irrelevant. In fact, if one side firmly believes that it has significant superiority (whether or not it is actually true), then the perceived balance is indeed relevant. A successful negotiator must be able to persuade the other side (as well as his own allies and sometimes even neutrals) that he believes in his own superiority. One purpose of procuring nuclear forces is to be able to make such a point in a bargaining session, and one purpose of a bargaining session is to explain why one side should back down and not the other. The influence and importance of courage, commitment, morale, and perceptions of strength have not changed in the nuclear era.

And there may still be rational reasons for going to war in the nuclear age. Not all nuclear wars will be accidental, inadvertent, or unintended, though such wars are, of course, possible. In the prenuclear era, accidental war was not a realistic possibility; today it is. In the nuclear age,

however, war may still result from many of the same causes for which
wars have always been fought:

1. For immediate national, ideological, or religious gains (where gains
    would outweigh possible costs, including consequences of mis-
    takes, bad luck, and critical uncertainties).
2. In the belief that long-range national, ideological, or religious pros-
    pects will be improved.
3. For abstract motives such as glory, plunder, boredom, power.
4. Under pressure to:
    A. Preempt a potential first strike by the enemy;
    B. Avert some other immediate disaster;
    C. Resolve some ongoing but increasingly desperate crisis;
    D. Avoid some long-range disaster by initiating a preventive war.

It is reasonably clear that the most likely reasons for a modern nuclear
war are 4A or 4B above; indeed, these possibilities have been much
discussed. Some observers also seem to believe that 4C and 4D are not
unlikely, but there seems to be widespread agreement that reasons 1 to
3 recede into the background as a cause of nuclear war (but might play
a critical role in tipping the balance when reasons listed in 4 were not
quite good enough by themselves). In that sense deterrence, especially
deterrence because of uncertainty, seems to work quite well.

As noted earlier, few nations are likely to go to war for positive gain
in the nuclear age. Wars are unlikely to be started deliberately, unless
they result from efforts to avoid what is perceived as an even greater
immediate disaster (for the Soviets, this might be loss of political control
over its East European empire; for the United States this might be the
losing of a conventional war in Western Europe). Nonetheless, in some
cases war might still be viewed and/or used as a continuation of politics
"by other means."

Although largely overlooked in modern defense planning, another
continuity between the prenuclear and nuclear era is the potential value
of a formal declaration of war. Such a declaration would make clear the
extreme seriousness of the issue at stake without necessarily producing
or requiring an immediate resort to armed conflict. But by keeping the
casus belli open it would also prevent immediate de-escalation of the
crisis. Once a war is formally declared, the crisis cannot be resolved
until a peace treaty of some sort is signed. The resolution of the conflict
remains pending until a response is made by the "wounded" party. A
formal declaration of war could be an extremely important device for

the U.S. if, for example, it were unable to respond effectively to serious aggression in a distant region. In such a case the U.S. would probably not want to initiate a nuclear war, but would want to declare its opposition and its determination to do something eventually. A formal declaration would give the U.S. time to mobilize its economy and its society.*

A declaration of war might be followed by a period of "phony wars" as happened in World War II. The U.S. and Soviet Union might, for example, have deployed their missile-carrying submarines, dispersed their strategic bomber forces, and placed their land-based missiles on high alert. An effective disarming counterforce attack would then be extremely difficult. And given the strength of the balance of terror and the fear of starting a major nuclear conflict, each side would be extremely cautious. They would seek to gain advantages through bargaining, threats, and harassment while avoiding a major confrontation that could easily lead to escalation. However, a "phony war" cannot last forever—either the crisis will be resolved (by negotiation or capitulation if one side is especially afraid of escalation) or it will escalate (to confrontation, as was the case during World War II).

A "calculated war" is another phenomenon that precedes the nuclear era and has not been conceptually affected by the development of nuclear weapons. A calculated war has at least four components:

1. An outbreak scenario (i.e., early stages of the war) that includes a variety of communication, bargaining, and tactical options that are dealt with rationally (even if the war starts irrationally)
2. If there is no rationality in the outbreak scenario, each side will still make rational attempts to protect its important values through:
   intra-war deterrence
   self-restraint
   counterforce
   active and passive defense
3. Either side would probably continue to bargain by:
   improving its threat position
   using abatement tactics
   special attacks and messages
4. A more or less rational attempt at war termination by use of what might be termed "end game" tactics—particularly cease fires and ultimatums

* No nation has declared war in the traditional manner since the end of WW II when the United Nations charter outlawed war. This being the case, a classic statement of intentions might be especially forceful.

In many ways the importance of the traditional options of warfare has not changed much in the nuclear era, except that many new problems have been added, particularly problems of timing and control. In the age of nuclear weapons, it is much more difficult to learn how to improve one's capabilities during a conflict, or to make the necessary corrections quickly enough to matter. The outcome of a nuclear war may well be determined in the first few hours, or on the first day, making the outbreak of a war much more important than it was in the prenuclear era (see chapter 6).

A final continuity between the prenuclear and nuclear era is the need for realistic military preparations. The ability of a country to mobilize for war, to protect and defend its people to the fullest extent possible, and to develop plans and capabilities for postwar recovery, remains an essential fact of national life. It can mean the difference between a "successful" war outcome and national disaster. Advocates of deterrence-only strategies often argue that preparedness in the nuclear era is irrelevant because nuclear weapons will destroy everyone and everything no matter what precautions are taken. As noted, I disagree.

The evacuation of civilians from potential target areas and the construction of an adequate system of shelters could substantially reduce the number of casualties caused by a nuclear attack on the United States. Civil defense is also important to enhance the credibility of our deterrent threat—the United States cannot threaten to attack urban targets in the Soviet Union (in retaliation for some Soviet aggression) if a Soviet counterretaliation will kill huge numbers of unprotected Americans. But if they have been evacuated and have a reasonable chance of survival, our deterrent threat becomes far more believable (see chapter 8).

Active defense, through the development of antiballistic missile systems, can also plan an important role in the protection of our military forces and (in the future) of our population centers. Defense against incoming missiles is difficult to achieve, and complete protection can probably never be guaranteed, but this does not make the effort any less valuable. If ten missiles are headed toward a city and eight can be destroyed before they explode, the level of damage done would be drastically reduced.

A final element of civil defense preparedness planning for postattack recovery—clearly a process that must be organized before a crisis occurs—should be designed to restore essential services as quickly as possible, including the ability to regain military strength if necessary. In a protracted nuclear war, where fighting would continue beyond the initial strikes, there would be a need for ongoing military production. Even with prior planning, recovery would require an enormous effort and probably take a fairly long time, but it could make a big difference.

In the nuclear as in the prenuclear era, the United States must still be able to fight and survive a war. It is not enough to take out insurance against nuclear war's occurring; the U.S. must also be able to stand up to the challenge of fighting. We must have a credible "alternative to peace," so long as the possibility of war—nuclear or conventional—remains. That possibility of war is one of the more important historical constants carried over into the nuclear era.

4

The Balance of Terror

Is the Balance of Terror Reliable? Is It Sufficient?

It is of first importance to both affirm and cast doubt on the widely accepted theory that the very existence of nuclear weapons in the U.S. and the U.S.S.R. creates a reliable balance of terror. Many people find the concept of "mutual homicide"—or mutual assured destruction (MAD)—very comforting; it strengthens the widely held (and hoped for) conviction that governments, once fully informed of the terrible consequences of a nuclear war, would never start one. The need to settle differences through other means becomes self-evident. The theory has other comforting aspects as well. For example, if destruction is inevitable, then expensive and complex preparations to reduce casualties, lessen damage, and facilitate postwar recovery are useless. We can spare ourselves the financial burden and—what is for many even more important—the intellectual difficulties, political problems, psychological discomforts, and moral ambiguities associated with such preparations. This is particularly true for some advocates of moral-pacifist, "deterrence-only," or "nuclear-weapons-are-just-another-weapon" positions.

According to the reasoning of most mutual homicide theorists, the destructive power of modern weapons is so enormous that only very few are needed to deter the enemy. Minimum or "finite" deterrence is sufficient. For example, in his 1979 State of the Union message, former President Carter seemed to imply that dissuasion through terror could be achieved with much smaller forces than in the past; larger forces would certainly not be needed:

> . . . just one of our relatively invulnerable Poseidon submarines—comprising less than 2 percent of our total nuclear force of submarines, aircraft, and land-based missiles—carries enough warheads to destroy every large and medium-sized city in the Soviet Union. Our deterrent is overwhelming, and I will sign no agreement unless our deterrence force will remain overwhelming.

But to believe that one survivable Poseidon submarine would constitute "an overwhelming deterrence force" is to calculate defense needs according to pre-1960 historical standards.

Minimum deterrence, or any deterrence predicated on an excessive emphasis on the inevitability of mutual homicide, is both misleading and dangerous. Deterrence should make a potential aggressor uncertain of success, and this fear can forestall an all-out nuclear attack even if neither side believes in the inevitability of Armageddon. Either side may doubt that the damage would be either total and/or inevitable but may still judge the risk to be unacceptable. One side may believe in the inevitability and totality of automatic annihilation much more than the other, and because of this may drop its guard (in which case the resulting negligence can be dangerous). But even if the belief is in fact mutual, many serious problems remain.

MAD principles can promote provocation—e.g., Munich-type blackmail or an attack on an ally. Hitler, for example, did not threaten to attack France or England—only Austria, Czechoslovakia, and Poland. It was the French and the British who finally had to threaten all-out war—a move they would not and could not have made if the notion of a balance of terror between themselves and Germany had been completely accepted. As it was, the British and French were most reluctant to go to war; from 1933 to 1939 Hitler exploited that reluctance. Both nations were terrified by the so-called "knockout blow," a German maneuver that would blanket their capitals with poison gas or unleash bombardment by conventional weapons. The paralyzing effect of this fear prevented them from going to war against Germany, and gave the Germans the freedom to march into the Ruhr, to form the Anschluss with Austria, to force the humiliating Munich appeasement (with the justification of "peace in our time"), and to take other aggressive actions.

A "balance of terror," therefore, may encourage an aggressor to feel that a guarantee by a guaranteeing power is unreliable or worthless. Aggression against the allies of the guaranteeing power may follow, or the allies may be discouraged from standing firm in a crisis. It is also conceivable (even likely) that an aggressor might conduct a limited war directly against the guaranteeing power since he is quite certain that the fear of further escalation will prevent the "limits" from being transgressed. As long as the aggressor feels safe from a large-scale or all-out attack, the fact that his opponent is also safe from such attack is simply not a sufficient deterrent.

On the other hand, if a potential aggressor himself fears escalation, he could judge the risks of an attack as being too high for the gains available. (This includes the concept of "deterrence against provocation.")

The deterrence value of the mutual annihilation theory is not unique to the West. Malenkov seems to have introduced it to the Soviet Union. Before him, the official Soviet position had been that war was inevitable; that capitalists—before accepting the defeat forced upon them by historical forces—would in a spasm of despair attack the bastion of socialism and have to be defeated militarily. Malenkov apparently argued that since nuclear war meant the end of civilization, the capitalists would not attack and the Soviets could therefore reduce their investment in heavy industry and military products and could concentrate more on consumer goods.

A different view seems to have been held by Khrushchev (at least initially) and by the Soviet military. They agreed that nuclear war would be disastrous and believed that the capitalist enemy might be deterred, but also might not be. They argued that nuclear weapons need not *necessarily* prevent war and that the Soviet Union could not, therefore, drop its guard. If the U.S.S.R. were sufficiently prepared in the event a war did occur, then only the capitalists would be destroyed. The Soviets would survive and, with luck, could create a worldwide postwar socialist community—a historical achievement that would more than justify whatever sacrifice and destruction had taken place.

This view seems to prevail in the Soviet military and the Politburo even to the present day. It is almost certain, despite several public denials, that Soviet military preparations are based on war-fighting rather than on deterrence-only concepts and doctrines—that is, on the belief that it is important to be able to wage and win (or at least survive) a nuclear war because a war may be forced on them (either by direct attack or by provocation leading them to strike first). They apparently believe that despite enormous damage they could win a nuclear confrontation, just as they "won" the Napoleonic invasion and World Wars I and II. In each case much of Russia was destroyed, but the country eventually emerged stronger than before.

Nonetheless, none of this implies that the Soviets would initiate a nuclear war except in the most desperate circumstances. One can imagine circumstances in which rational leaders rationally decide to initiate a thermonuclear war; in particular, to risk this course as the least undesirable available option. Alternatively, a nuclear war might result from an accident, miscalculation, irresponsible behavior, or other inadvertence (especially in a tense situation).

The balance of terror is sufficiently effective that it is difficult to describe plausible circumstances in which a leader would decide that an all-out war,* or even a limited war with an appreciable risk of further escalation, was a desirable alternative. It is even difficult (but not impossible) to formulate "not implausible" scenarios in which war is the least undesirable alternative.

Assessing the Feasibility of Nuclear War

To support the assertion that a "rational" judgment about the "feasibility" of thermonuclear war even exists, and to support the claim that it would not be fought purely or mostly for positive gains but out of desperation, one must analyze the kinds of gains and costs (including largely unknown risks) that decision makers would have to weigh. Their deliberations over the feasibility of war greatly affect peacetime preparations—particularly the allocation of resources to various war-fighting versus deterrence-only strategies.

Among the determinations U.S. and Soviet political and military leaders would have to make are those related to military programs. These include:

1. The various phased programs (including "mobilization bases") of both sides for deterrence and defense;

---

* "All-out" in the sense that no forces are held in reserve except for military purposes, but the targeting—at least initially—could still be quite controlled and limited.

2. the wartime performance of U.S. and Soviet forces under different preattack and attack conditions (including a war of multiple strikes occurring over a period of weeks or months);

3. the known consequences of all relevant weapons effects (e.g., nuclear and thermal radiation, electromagnetic pulse, dust, ground shock);

4. attacks designed to exploit or diminish these effects for narrow military purposes, for coercion, or for intrawar deterrence;

5. other intrawar contingencies (e.g., if the war were one of extended duration, intrawar reconstitution of military forces of both sides could be important); and

6. the characteristics of war termination (recognizing that even all-out thermonuclear wars come in different shapes and sizes and can have greatly varying outcomes).

Decision makers also would need to be concerned with the anticipated postattack problems of short-run survival and reorganization ("patchup" operations as well as the restoration and maintenance of economic momentum), and the prospects for long-term recuperation. Clearly, all of these would largely depend on the degree and nature of the nuclear damage. The effectiveness of any active and passive defenses in mitigating the damage would be another important factor, and the course of recovery would also be affected by the political outcome. How much of the government apparatus would survive a war and be available to aid in recovery efforts? Would recovery proceed in a "benign" political environment or a hostile one where repeated enemy strikes or threats would inhibit the work of restoration?

Long-term recuperation would be influenced by the domestic and worldwide political, economic, military, and social context that existed in the aftermath of a nuclear war. External trade and other aid might be crucial to long-term recovery. Military rearmament might be necessary.

Finally, some uncertain and potentially catastrophic environmental aftereffects (a new Ice Age; destruction of the ozone layer) would undoubtedly influence the decisions of U.S. and Soviet leaders. These and other unknowns are among the reasons for the validity of the assertion that neither side would go to war frivolously, happily, or for "purely or even mostly positive gains."

The Soviets could start a war in any crisis situation where they perceived the risk of *not* going to war to be greater than the risk of *going* to war. Such circumstances resulted in Soviet suppression by military force (directly or through a Soviet-dominated local government) of several revolts in Eastern Europe (East Germany, 1953; Hungary, 1956; Czechoslovakia, 1968; Poland, 1981). In each case, the United States felt some pressure to intervene but chose not to; in fact, it made concerted efforts not to "rock the boat." However, assume the United States or its West European allies were to intervene directly in a future satellite uprising. The intervention would be less likely to involve open military operations against Warsaw Pact forces than to involve the provision of arms to the rebels or vociferous diplomatic support for their cause. In any such crisis, the Soviets might be faced with four agonizing choices:

1. They could do nothing. This would be dangerous. The "afflicted" satellite might become permanently independent of Soviet dictates, and the uprising could spread then or later to other countries in Eastern Europe or even to the Soviet Union itself. In any case, Soviet prestige, both domestic and foreign, would be damaged greatly; its military control in East Germany might also be endangered (e.g., if a Polish revolt endangered their logistics and communications).

2. They might resort to some limited, temporizing measures: negotiation, a conference, a change in policy or government, some kind of fait accompli (e.g., an assassination or a coup d'etat) which, if successful, would greatly alleviate the problem. But these measures could easily be ineffective or backfire.

3. They could intervene with Soviet forces (as in Hungary in 1956, Czechoslovakia in 1968, and Afghanistan in 1979), but these interventions were political disasters and not always immediately effective either. The Soviets might fear an expansion of U.S./NATO limited intervention, perhaps a renewal of the cold war including a massive mobilization, or even more drastic action by the U.S.

4. Rather than wait and risk further escalation, or risk a later showdown of some kind at a time chosen by the Americans, the Soviets might decide to hit the United States right away, thus attaining the all-important first-strike advantage.

This kind of potentially dangerous situation is not likely, but it cannot be ruled out. It could happen in an East German revolt in which frontiers were opened up by the rebels and West Germany was gradually dragged into the fight. In that case the U.S. retaliatory capability would have to be so impressive that the Soviets would believe a first strike entailed greater risks than any other alternative (even after a Soviet evacuation of its cities). For the U.S. to achieve a genuinely credible retaliatory capability (i.e., one that could effectively deter under great strain and do so with the necessary assurance) in the near future would be harder to accomplish than many suppose, although current capabilities are more than marginal.

How Satisfactory Is a Reliable Balance of Terror?

Even if one accepts the balance-of-terror theory, including the belief that there are almost no circumstances in which the Soviets would launch a deliberate attack on the continental United States (and vice versa), some important strategic problems remain. For example, the Soviets might attack U.S. "vital interests" (e.g., Western Europe, or U.S. overseas forces) instead of the United States itself. Would the United States retaliate by an all-out attack on Soviet cities? It seems very doubtful. In fact, a thermonuclear balance of terror is functionally equivalent to a nonaggression pact: neither the Soviets nor the Americans will initiate an all-out attack on the other, no matter how provocative the behavior—short of a major attack on one's homeland. This concept of American reluctance is not new. Starting with the assertions of Secretary of State Dean Acheson and Under Secretary Christian Herter that the U.S. would distinguish between attacks on its allies and attacks on itself, there has been concern over the possibility of U.S. inaction in the wake of a Soviet attack on Europe. This concern was a primary motivation for British and French acquisition of independent nuclear deterrents, for the discussion of the "limited nuclear options" suggested by Secretary of Defense Schlesinger in the early 1970s, and for the NATO decision to deploy Pershing II ballistic missiles and GLCMs (ground-launched cruise missiles).*

Most Americans agree that any president would exercise restraint rather than act in a rash or self-destructive fashion. They believe his first concern would be for the future of his country, then for its allies, and then for the world—most likely in that order. Until about 1970, many Americans argued that a U.S. president might well risk the lives

---

* While there is now much West European opposition to such deployment, the initial impulse came from West Europeans who wished to make the U.S. nuclear guarantee more effective by ensuring an almost automatic escalation (via the Pershings and GLCMs) to a U.S.-Soviet exchange following a Soviet attack on Western Europe.

of as many as 50 or 60 million people for the Western alliance. Few feel this way today. While there is a strong feeling that the resolve of America to honor its NATO commitments has lessened, and that U.S. ties with Europe are less strong and emotional than they were twenty or thirty years ago, it is nonetheless clear that the United States would still risk a lot for its allies.

When Europeans are asked how many casualties they believe Americans would be willing to suffer to live up to their NATO obligations, their estimates, perhaps not surprisingly, are much lower: they range from 0 to 10 million. Some distinguished European experts argue that the United States would not retaliate against a major Soviet aggression in Europe if the Soviets threatened to destroy only five to ten empty U.S. cities in a counterattack. But almost all Europeans agree (after some hesitation) that Europeans would not do anything similar (i.e., risk even partial self-destruction) for the United States. The U.S. alliance with Europe is mostly a one-way and probably a deteriorating relationship.

Do the Soviets find the threat of U.S. retaliation credible? We know that Soviet decision makers are strongly conditioned to control their emotions so that an immediate or obvious (but ultimately self-destructive or counterproductive) response would be frowned upon. Soviet leaders seem to believe that one should not be provoked into self-destructive behavior and probably assume that Americans think the same.

But this reasoning could be wrong. There is no American tradition of controlled reactions to assaults on "national honor" or American "principles." Indeed, the opposite has been true—as in "Give me liberty or give me death!" If the Soviets are familiar with American history and are cautious, they will realize this.

However, if the Soviets were to test U.S. resolve by instigating a series of crises, they would learn a great deal about how much provocation the United States would take. But such "testing" would likely influence or change U.S. attitudes and policies rather than provide reliable or useful information. (It is a commonplace of history that attitudes may be dramatically altered by events.) But the real problem in a crisis could be to convince the Europeans to stand firm, i.e., to get them to believe that there will not—and should not—be excessive accommodation, appeasement, or surrender; that the choice is not "everybody Red, everybody dead, everybody neutral" or some similar unpalatable set of alternatives, but "everybody Red, everybody dead, or everybody NATO."

Official U.S. Government unclassified estimates of U.S. casualties in a large-scale nuclear war run from 20 to 150 million or more. I believe that with proper tactics and some quick fixes, both the lower and upper ends of the range given would be much lower. For example, if the United States, on receipt of "unambiguous" strategic warning, were to evacuate its cities, shelter its population, and launch a preemptive (not preventive) attack on Soviet counterforce targets. If the official estimates are relevant, however, then the U.S. would almost automatically appear to be precluded from honoring any strategic guarantees. Under the circumstances, deterrence becomes more important than ever.

5

Three Kinds of Deterrence

Deterrence is not a single, or simple, concept. It is important to distinguish between three different and especially significant types of deterrence. "Type I" is strategic deterrence of a massive direct attack against one's country or a large portion of its forces. The deterrent is the threat of a major retaliation. "Type II" is deterrence against extremely provocative acts other than a massive direct attack against one's country or its major forces (e.g., an attack, or threat of attack, on an ally). Strategic threats provide the deterrent. "Type III" is tit-for-tat deterrence against limited military or nonmilitary actions. The deterrent is the fear that some kind of proportionate retaliation will make the aggression unprofitable.* All three types of deterrence need some amplification.

* For those who prefer not to use numbers in this way, I suggest the labels "Basic (Strategic) Deterrence," "Extended (Strategic) Deterrence," and "Provocation Deterrence by Limited Means," or some shorthand versions of these phrases. The word "strategic" in the first two terms indicates that the type of deterrence entails the use of strategic nuclear threats. Both in terms of the provocative act and the counterthreat, many other kinds of deterrence could be discussed.

Type I: Deterrence Against a Direct Attack

Almost everyone agrees that Type I Deterrence must be made to work; we simply cannot risk the possibility of failure. Never have the stakes on prevention (as opposed to correction) been so high. But the extreme view—that deterrence is everything and all else is hopeless or counter-productive—is questionable, as is the idea that there should be no trade-offs with other objectives.

In the 1950s and early 1960s, most discussions of the conditions for Type I Deterrence tended to be unrealistic. Typically, the debate focused on comparing the preattack inventory of U.S. forces versus Soviet forces, comparing the number of bombers, ballistic missiles, and submarines on each side. This was basically a World War I and World War II approach; unfortunately much of the current public discussion is very similar.

What had become moderately well understood, even in public discussion, was that the essential numbers are the estimates of the damage that strategic forces could inflict *after* being hit. Much recent discussion, however (especially concerning the "window of vulnerability"), leaves this postattack calculation out and reverts to inventory calculations. In fact, an evaluation of a nation's strategic capacity must take the following contingencies into account:

1. That the attacker could strike at a time and with tactics of his own choosing;
2. that the defender might have to strike back with a damaged and perhaps uncoordinated force in a postattack environment which has degraded his ability to counterattack, possibly in unexpected ways;
3. that the attacker's defense force might be completely alerted and his cities at least partially evacuated;
4. that the attacker might use blackmail or intimidation to condition the defender's response—i.e., that there could be intrawar deterrence as well as prewar deterrence. The attacker could employ measures that enhance intrawar deterrence, and even if such measures work only for a short time, they might be quite effective in limiting damage to the attacker.*

* Recall my exchange with Senator Proxmire on the possibilities for intrawar deterrence, pp. 70–71.

The first step in this calculation—an analysis of the effects of a strike on the defender's retaliatory ability—depends critically on the attacker's tactics as much as on his capabilities. For some issues, the question of warning is uppermost (e.g., for ensuring the survivability of vulnerable ballistic missiles, strategic bombers, and airborne command-and-control aircraft). But the enemy may degrade the defender's force effectiveness in other ways as well, for example, by interfering with command-and-control arrangements.

Thus, in evaluating an enemy's capabilities, it is important to look beyond the conventional tactics that the standard assumptions lead one to expect, since a clever enemy might employ creative and unconventional methods. A defender should not assume what Albert Wohlstetter has called "defender-preferred attacks," i.e., those a potential defender feels most able to deal with and therefore would prefer. Instead, the focus should be on "attacker-preferred attacks," namely those a desperate or highly ideological aggressor may prefer. Attacker-preferred scenarios should factor in the attacker's defense doctrine and "strategic culture," not only as presented in its official military literature, but also those alternatives that may be more sensible, rational, creative, and ingenious than the ones discussed in print.* War planners often come up with very imaginative and unexpected ideas—particularly when the penalty for not doing so becomes both great and immediate.

The definition of strategic force "survivability" used until quite recently by most U.S. defense officials illustrates the danger of taking insufficient account of attacker-preferred attacks. America's efforts to improve its force survivability traditionally emphasized the ability of ICBMs, bombers, and strategic submarines to withstand or evade attacks intended to disarm. Relatively little attention was given to ensuring the survivability of the command, control, and communications (C^3) networks needed to direct these forces during an attack. As a result,

---

* Richard Betts has documented numerous historical cases in which attackers weakened their opponents' defenses through the employment of unanticipated tactics. These include: rapid changes in tactics per se, false alarms and fluctuating preparations for war, counter-intuitive leapfrogging on the "escalation ladder," creation of spurious indicators of military activity, and the combination of technical and doctrinal innovations to gain surprise. (See "Surprise Despite Warning: Why Sudden Attacks Succeed," *Political Science Quarterly,* Winter 1980–81, pp. 551–572.) This is exactly the kind of thing which is likely to surprise those who subscribe to MAD theories. Those who see a need for warfighting capabilities expect the other side to try to be creative and use tactical innovations such as coercion and blackmail, technological surprises, or clever attacks on "leverage" targets such as command-and-control installations. If he is to adhere to a total reliance on MAD, the MADvocate has to ignore these possibilities.

Soviet strikes on American C^3 assets might be as deleterious as an attack on the strategic forces themselves (the "defender-preferred attack").

According to what we know about Soviet preparations and military discussions, the Soviets appear to be much more cognizant of the problem of C^3 vulnerabilities than we are. In addition, there are unanswered questions about the wartime performance of command-and-control systems: for example, the electromagnetic pulse (EMP) generated by the high altitude detonation of only a few multimegaton nuclear weapons over the United States could damage almost all inadequately protected electrical and electronic equipment, or could do little damage. The effects of an EMP attack are simply not well understood, but the Soviets seem to know—or think they know—more than we do. In any case, the problem of operating in a postattack environment after training in a peacetime environment can be compared to training with a delicate system at the equator, damaging the system, and then moving to the Arctic where the (now damaged and imperfect) system is expected to work efficiently the first time it is tried.

The second factor in the calculation—coordination of the defender's surviving forces—depends greatly on the defender's own tactics and flexibility. If, for example, the defender's offensive force is spread thinly over a large target system, and if the attacker succeeds in destroying much of it, many important targets of the defender's retaliatory strike would go unattacked. If, on the other hand, the defender concentrated force assignments only on important targets, he would probably waste much of his force strength by overkill of a few high-priority objectives.

For many reasons, therefore, it would be wise to evaluate the damage before retaliation, but that may not always be possible. Much would depend on the timing of the attack, the nature of the targeting process, and the poststrike capability for evaluation, command, and control. In addition, the rapidity of changes in the technology of war makes it especially difficult for the defender to react quickly and effectively to technological changes in the offense.

In addition to the physical vulnerability of its forces there is also the psychological vulnerability of a nation's resolve. Imagine, for example, that the United States' fleet of ballistic missile submarines, which is invulnerable to an all-out simultaneous Soviet attack, is subjected to Soviet destruction one vessel at a time at sea. The U.S. president would have to stop this attrition. But assume that he does not possess a "not-incredible counterforce first-strike capability." Thus, if he gave the Soviets an ultimatum, they might effectively preempt a U.S. first strike; if he attacked Soviet cities, the Soviets could do even more damage in retaliation than was done to them by the initial strike. The president simply could not risk the ultimatum, much less an attack.

There are many other ways either side can make post-attack coercion attack feasible, such as holding enemy cities hostage for the purpose of intrawar bargaining. These tactics and threats would involve problems of timing, control, communication and persuasion, but the problems need not be insurmountable. While the attacker probably could not be sure that these tactics would work, he might be willing to try them, especially if he had decided to go to war anyway. The extra cost might be small and the potential gains very large.

One of the most important, neglected, and uncertain elements of many American retaliatory calculations is the effect of Soviet civil-defense measures. Again, until recent years, the Soviets were seldom credited with even modestly effective preparedness; some Western analysts went so far as to assume that the Soviets would take no precautions to protect their population in case of war. A much more reasonable alternative was almost never realistically considered: that in order to reduce the vulnerability of their own population, the Soviets might evacuate their cities to existing or improvised fallout protection.

Some analysts believe that since the deterring nation strikes second, civil-defense preparedness could reduce the effectiveness of Type I Deterrence. It could make a purely military attack more likely by increasing the possibilities for intrawar deterrence, postattack blackmail, and a "reciprocal fear of preemptive attack." Other analysts argue that even if this were true, civil-defense measures are justified by the great reduction in damage if war occurs. They believe extensive civil-defense preparedness could enhance war-fighting capabilities to the point of actually improving Type I Deterrence against military attacks. Both sides have valid arguments; civil defense can go either way (see chapter 8).

The effects of a first strike by missiles are much more calculable than almost any other military operation; in principle, one need only apply well-known principles of engineering and physics and use relatively certain data and assumptions for an analysis of the immediate impact of the attack. But the fact is that the results of attacks are not mathematically predictable, even with tested missiles. The probability of extreme variations in performance, the upper and lower limits, and especially many of the effects of the weapons, cannot be calculated accurately. But many laymen (and too many professionals) continue to regard the matter as a simple problem in engineering and physics. A Soviet general might be able to convince his audience that the Soviets would suffer very little damage from firing ICBMs, and certainly not as much damage as they suffered in World War II. Unless sophisticated objections were raised concerning the possibilities of intelligence leaks, firing-discipline problems, unreliable basic data, field degradation, and so on, the general's argument would be quite plausible. But if somebody did raise those objections, there are several caveats the Soviet general would have to concede.

Given the relatively small group involved and the tight security within the U.S.S.R., avoidance of leaks on any significant scale could probably be accomplished fairly easily. But what about the possibility of a very senior or critically placed Soviet defector—possibly another Colonel Penkovsky?* And while the firing discipline of an attack can probably be controlled through proper doctrine, operational training and testing, and various technical devices, these may not be reliable if the orders to fire come out of the blue (i.e., if the system has not been alerted in advance). On the other hand, if there is a preliminary alert, the chance of a leak is increased.

The basic data are crucial—the Soviets must have accurate intelligence about the U.S. military posture. But even more significant, and not necessarily easy to come by, is accurate information on their own military posture—the actual yield, accuracy, and reliability of their ICBMs, for example. It is surprisingly difficult to obtain truly reliable estimates of these quantities. Most important, the system will never have been fully tested under realistic conditions. If the Soviets have a much better technological capability than they need, then they don't need especially reliable estimates. But if their capability is only marginal, they had better know exactly how good it is.

The caveat on field degradation is obvious: what happens when the missiles are actually operated in a dispersed mode by regular military personnel? The Soviets do more field testing than we do and still the worry remains—how far off from range performance and maintenance will real performance be? Given the historical record of Soviet incompetence and the recent lack of reassuring experience on military estimates (e.g., Afghanistan), this issue should raise anxiety.

* Colonel Penkovsky was a high Soviet official with access to Soviet secrets who delivered an extraordinary amount of highly classified material to the Central Intelligence Agency in the 1960s before being found out and executed.

The Soviets could hedge their bets as to how successful a first strike would be by using intrawar deterrence and/or postattack blackmail. If they concentrated their attack solely against strategic targets (and airburst some of their larger warheads to minimize local fallout), then they might be able to limit U.S. casualties (to a few million) and claim that they had withheld enough forces to totally *destroy* our country and did we really want to pick this moment to initiate the use of nuclear weapons against open cities?

The United States has taken many measures to alleviate problems connected with Type I Deterrence, but the primary need remains: to be able to react rapidly even to hypothetical changes in the enemy's posture. (Many people in peace groups and/or advocates of deterrence-only positions don't want to worry about such details.) The United States will have to continue to update its defense posture for the indefinite future if it expects to be able to deal with situations in which the Soviets are under great stress and to deal with them with some assurance.

This last point is important. When decision makers and the general public evaluate the quality of our Type I Deterrence, they usually ask whether it is strong enough to prevent the Soviets from attacking us in cold blood in normal situations. Since almost any minimal nuclear deterrent force would suffice for this purpose, this approach is misleading. It is more appropriate to evaluate the quality of Type I Deterrence by asking how much strain it could accept and still be dependable. At what point—and there presumably has to be such a point—is the strain likely to be excessive? The following discussion of Type II Deterrence suggests some plausible circumstances in which Type I Deterrence could be strained, and our estimate of how far we can afford to strain it; this is the central issue.

Type II: Deterrence of Extreme Provocations by Threat of Strategic Attack

Type II Deterrence is based on a different calculation from Type I; nevertheless, it is still a Soviet calculation. The Soviet decision to proceed with a very provocative move will be influenced by their assessment of the American response. For example, would the Americans strike first* and damage Soviet strategic forces (after which the Soviets

---

* Legally, morally, and politically, the U.S. might be retaliating against a Soviet provocation (such as an attack on Europe), but from a strategic viewpoint, it would be a first strike.

would have to strike back, uncoordinated, and possibly against an evacuated U.S. population)? If they conclude that the U.S. might attack, then they also admit that their own Type I Deterrence could fail. If they believe their Type I Deterrence is overwhelming, then they may go ahead with the provocation.

U.S. Type II Deterrence involves the possibility of a U.S. first strike, perhaps after making some temporizing move, such as evacuation. (A U.S. evacuation becomes plausible if the Soviets are, say, invading Western Europe, but are trying to keep the war limited to Western Europe.) This first strike presupposes adequate strategic defenses (civil, air, ballistic missile) to back the threat. However, most U.S. analyses of proposed active and passive defense programs assume a Soviet surprise attack directed primarily against U.S. civilians; this is an inappropriate (and unnecessarily pessimistic) context for evaluating the contribution of such defense preparations to Type II Deterrence. A U.S. first strike—after extreme Soviet provocation—would be more relevant in evaluating the role of U.S. offenses and defenses in Type II Deterrence.

Even a moderate nonmilitary U.S. defense program is likely to look impressive to the Soviets (and probably to most Europeans). For example, the Type II Deterrence scenario provides a solution to the crucial problem of obtaining adequate warning. The Soviets are likely to think that we have more freedom than we do, and that we are not worried by the possibility of giving them strategic or tactical warning (though in fact planning a retaliatory, i.e., strategic, first strike could involve a considerable risk of the Soviets getting advance warning). U.S. planning and actions would therefore have to be tempered by the realization that a disclosure (such as might result from a U.S. evacuation) or a mistake could trigger a preemptive Soviet attack.

Augmentation of our active and passive defense system could also give us a nonattack option in the face of Soviet provocation. For example, after evacuating our cities to fallout-protected areas and alerting and deploying various strategic forces, the U.S. could emphasize the strength of its position should it choose to initiate hostilities. The U.S. capability need not necessarily be "credible"—only "not incredible." Under the circumstances, the Soviets basically would have four broad alternatives:

1. To initiate some kind of strike against U.S. forces;
2. to prolong the crisis, even though it would become increasingly credible to expect a U.S. first strike if they continued to do so;

3. to carry out some decisive action that more or less terminates the old crisis (but may result in a new one);
4. to back down or compromise.

Ideally, the Soviets would select the fourth alternative because the American Type I and Type II Deterrence would, respectively, make the first two choices sufficiently unattractive, and either its Type II or Type III Deterrence could make the third choice so unattractive as to leave only option 4.

Mulтisтable Deтеrrence

Before going on to Type III Deterrence, it is important to recognize the value of combining I and II above, even though all of them are some-what ambiguous and blend into each other. The concept of multistable deterrence implies that it is possible for two opponents to possess, simultaneously, reasonably satisfactory levels of Type I and Type II Deterrence. This seems paradoxical, because to some extent, one side's Type I Deterrence is measured by the inadequacy of the other side's Type II Deterrence. A partial resolution of the paradox lies in the fact that nations tend to be conservative and tend to look at the worst scenarios that might reasonably happen to them. Thus, the calculations made by both sides will be inconsistent because both sides will hedge. To the extent that nations tend to make optimistic calculations, it is more difficult to have multistable deterrence.

The basic characteristic of a situation in which there is multistable deterrence is that both sides have a good deal of Type I Deterrence, but in addition, they have an ability to threaten a nuclear attack in order to deter extreme challenges to their existence. Deterrence against a threat of attack is very high but is not absolute: both sides have a very high deterrent against striking the other side but not so high they can't be provoked; and there is an advantage in attacking first. But by denying either side victory and by threatening the other side with great damage, a first attack is discouraged, and neither side will feel it has a free ride to be as provocative as it wants. It is stable against attack and stable against provocation. In addition, it should also be stable against accidents. Where multistable deterrence exists, the threat of a calculated nuclear attack will serve to constrain the political conduct of both sides.

While this definition has been framed in terms of the *size* of possible retaliatory blows, essentially the same situation can be achieved on the basis of the *probability* of very large retaliation. That is, multistable deterrence could exist if each side (a) had a 50 percent chance of delivering an overwhelming retaliatory blow if it were attacked; and (b) had a 50 percent chance of escaping without overwhelming damage if it made a first strike against the other side. In practice, multistable deterrence will present a mixture of the quantitative and the probabilistic factors. It may be somewhat unstable from the point of view of arms-race considerations, in that it could be relatively sensitive to technological and force changes.

On the other hand, it is important to note that a deterrent situation that is very stable against preemption or first strike may, in some sense, actually encourage extreme provocations. A situation in which there is multistable deterrence, although it is somewhat less stable against surprise attacks and unintended war, provides greater stability against provocations; that is, provocations do not increase because of a lesser danger of central war. Essentially, then, multistable deterrence has a considerable capability to alleviate the consequences if stability fails and if war or provocation should occur.

The situation has two major instabilities, however: (1) it encourages an arms race, since improvements in numbers and quality are of significance; (2) as noted above, in most postures that do not involve automatic mutual annihilation there will be an advantage in striking first. (There is, however, no logical necessity that this be true. One could imagine a situation in which the attacker used up more resources than he destroyed, but I think of such a situation as being technologically improbable unless one or both sides have deliberately accepted weak postures or tactical restraints.)

Type I Deterrence depends upon survivable nuclear forces; the credibility of Type II Deterrence depends upon the ability of the U.S. to limit damage to itself by some combination of intrawar deterrence and counterforce nuclear strikes to disarm the enemy. Nuclear-powered, ballistic missile submarines, because of their presumed capability to survive a strike and retaliation, are commonly believed to be one of the most effective components of Type I Deterrence force. Accurate MIRVed ICBMs are commonly believed to be one of the most effective Type II Deterrents.

When a nation possesses a significant quantity of both Type I and II Deterrence, i.e., when it has multistable deterrence, it is relatively secure against both nuclear attacks and extreme provocation short of nuclear attack. The concept of multistable deterrence is very important in the NATO context. West Germany is prohibited from possessing either type of nuclear deterrent force. Great Britain and France have relatively weak Type I forces with some latent Type II capability but not much. Thus, the U.S. Type II Deterrence substitutes for the general lack of European Type I Deterrence. This relationship obviously involves greatly differing types of risks. The Europeans depend for a significant portion of their security upon the U.S. The Americans protect Europe from nuclear attack by, among other measures, making it more likely that their own homeland will be attacked. Unless U.S. Type II Deterrence is credible, American nuclear guarantees to Europe lose their salutory effect. This complicated relationship makes arms control between the U.S. and the U.S.S.R. difficult to attain. At the same time, a sound understanding of this relationship in the West provides a sound basis from which to negotiate in those situations where arms control is pursued.

To follow this line of analysis further, it is perhaps more accurate to discuss "symmetrical multistable deterrence." This is to say that to a large extent, both the U.S. and the Soviet Union have multistable deterrent forces that exist in a sort of symmetry. (The question of whether, or to what extent, this is in fact the case today will not be addressed here.) The following table lists the factors involved in multistable deterrence:

SYMMETRICAL MULTISTABLE DETERRENCE

1. Both sides have a significant first strike advantage (Type II Deterrence)
2. Both sides have sufficient second-strike capability (Type I Deterrence)
3. Thus neither side wants war, but if war is "inevitable" they must prefer going first
4. If war isn't inevitable, then threat of "3" above is enough to deter extreme crisis, without being very destabilizing in ordinary crises.
5. Thus preemptive or other war might or might not be more likely while there was an extreme crisis, but extreme crises are less likely and more easily resolved. This last, on balance, probably improves stability. Either a one-sided or mutual recognition of this could further reduce the probability of preemptive war.

As one can see from the above table, symmetrical multistable deterrence mutually deters various forms of war and extreme provocation by relying on what appears to be a paradoxical formulation—by making nuclear war marginally more likely in a serious crisis, one actually makes all forms of war—including nuclear war—less likely. The key to understanding this formulation lies in the following chain of reasoning:

1. Symmetrical multistable deterrent forces produce a reciprocal fear of surprise attack in a crisis;
2. this fear is intentional—by design, not by accident;
3. as a result, extreme provocations will be less, not more, frequent;
4. since nuclear wars are most likely to start as a result of extreme provocation (rather than out of the blue), they are less likely if such provocation is deterred.

Stated in this fashion, the concept of multipolarity is almost certainly an oversimplification of reality (whatever that may turn out to be). Several qualifications need to be added, therefore, though we do not feel that they negate the essential validity of the symmetrical formulation as stated above. First, the distinction between abstract calculations and real political-military decisions is crucial. In particular, statistically controllable uncertainties in the realm of the former become enormous doubts in the realm of the latter. The difference between the analyst and the politician can be illustrated by the following: The politician may simply comment, "Our forces can do unacceptable damage to the opponent and therefore he won't challenge us." The analyst, to paraphrase a remark by Raymond Aron, may look at deterrence in the following way: who deters whom, from what acts, in the face of what threats (themselves facing what counterthreats), in what context, for what purposes? These uncertainties involve capabilities (i.e., the ability to destroy hardened targets) and enemy responses. (I.e., if I exercise restraint, will he?)

Political leaders may not see the same clear distinction between Type I and II Deterrence that military planners see. (In some ways, the opposite is at least as likely.) In either case, these uncertainties are not unique to nuclear war. The fact that we have never experienced this phenomenon does not create the doubts—all wars are plagued with uncertainty and always have been. The stakes may be higher in a nuclear war, but leaders in the past have risked their country's very existence on a military gamble.

We should note that the responsible decision maker is likely to be somewhere between the analyst and the politician. We expect the driver of a bus to be more careful than the driver of a car; the driver of a school bus to be more careful than the driver of a regular bus; the engineer on a train to be more careful still; and the pilot of a plane to be even more careful. The president, who is the leader of the free world and responsible for his constituencies, simply cannot take either the casual view or the overly technical expert's view. While he is not going to be as preoccupied with details as the analyst, he is going to be much more concerned with these than even a responsible and prudent average citizen.

Under most so-called "normal" (i.e., noncrisis) circumstances, most decision makers will resort to Type III Deterrence much more frequently than to Type I, II, or multistability. What, then, is Type III?

Type III: Deterrence of "Moderate" Provocations

There is a range of hostile relationships—from on-going, low-key hostility to relatively high but still limited levels of violence—that require a third kind of deterrence against "moderately provocative" acts. The hostility is influenced by the strategic balance (real or perceived) and by the respective strength of each side's Type I and Type II Deterrence, but they are affected primarily by capabilities that are immediate, directly credible, and "usable." These include:

1. Moral, political, legal, and ethical reprisals and constraints;
2. various sanctions and other consequences of breaking a custom or crossing a threshold;
3. alienating friends or antagonizing neutrals;
4. creating or strengthening hostile coalitions;
5. increasing the military capability of specific potential opponents —e.g., actuating what we call "mobilization bases";
6. immediate escalation or "excessive" countermeasures.

Numbers 4 and 5 are especially important: if the provocation creates a sense of danger among potential enemies, the enemies may start to organize (individually and collectively) to create some effective threats and defenses. Even in a thermonuclear era, any continued, dramatic, or repeated provocation is likely to result in the creation of countervailing forces. And unless the aggressor is or becomes extremely strong, these countervailing forces will eventually prevail, subjecting the aggressor to some unpleasant risks in the process.

Countervailing forces are difficult to control once they achieve a certain momentum. Indeed, their development represents one of the classic methods of foiling aggressive movements. But in the thermonuclear age this might not work because of the small credibility of the threat to actually *use* (instead of just deploy or mobilize) the highest level of force.

In the early days of the nuclear era, many civilian strategists felt that the balance of terror was so reliable that even an extremely hostile act would not provoke escalation to general nuclear war. Today the (more accurate) assumption is that the nuclear nations probably are reluctant to risk upsetting the status quo, and there is little that would make them do so. To even remotely risk an Armageddon for immediate gains or potentially favorable longer-term scenarios is almost unthinkable. As a result, deterrence works. But its effectiveness depends heavily on the weakness of the desire to be aggressive, as well as on the maintenance of competent military establishments.

6

Some Scenarios for the Outbreak of Nuclear War

The Implausibility of Nuclear War

Military planning for nuclear war is greatly complicated by the difficulties one encounters in conceiving of plausible circumstances for its outbreak. Even defense analysts who recognize the necessity of planning to fight a nuclear war confront this problem. The difficulty lies mostly in two developments: (1) deterrence seems to be working; and (2) world leaders have a basic disinclination to go to war, or even to run great risks.

For the first time in history, external aggression is one of the least effective ways for a nation to increase its wealth, power, influence, or prestige. In fact, the pressures against war, and especially against nuclear war, tend to be overwhelming. Today, even quasi-ideological movements, such as Soviet communism, do not generally think of expanding their influence through naked military (especially nuclear) aggression. Armed force, if used at all, is used to support or reinforce a previous political gain (which itself may have been obtained by the use of low-level violence combined with propaganda and agitation). Alternatively, armed force is used to stave off political defeat—as in Hungary (1956), Czechoslovakia (1968), Afghanistan (1979) and Poland (1981).

While the above is not a universal rule of contemporary defense policy, it is close to one; close enough so that almost no military and civilian analysts can create truly plausible scenarios for the initiation of a nuclear war. Even though I do not believe in total mutual assured destruction, much less in any automatic "overkill," these concepts create such a strong deterrent effect that the concepts themselves might as well be true for most peacetime and even crisis circumstances.

Almost all analysts believe that the enormous prospective risks of a central nuclear war—even one that did not escalate to mutual annihilation—would normally prevent decision makers from considering it a practicable option. They can hardly imagine even an irrational decision maker (let alone a rational one) starting a nuclear war. And if a leader did become irrational and tried to start a nuclear war, they believe that others within his government or country would probably stop him.

The decision not to go to nuclear war is essentially automatic; it is really a nondecision. Because the resort to nuclear use is not considered a serious option in the 1980s, many factors that could influence the outcome of a nuclear war (e.g., wartime tactics, the performance of nuclear forces in combat, civil defense preparations) are not considered seriously either.

The implausibility of nuclear war was less pronounced in the 1950s. At that time the circumstances, if not always the specifics, of the outbreak of a nuclear war were easier to imagine. The implications of a nuclear conflict—even after Hiroshima and Nagasaki—were not yet as awesome (the explosive power of most weapons was still measured in kilotons, not megatons); Stalin (and later Khrushchev) was thought of as a Hitler-type gambler who might resort to anything, including nuclear war; and the U.S. nuclear superiority was so overwhelming that American use of nuclear force to counter a Soviet provocation (such as a conventional attack on Western Europe) was very plausible. In fact, the U.S. used both implicit and explicit nuclear threats to constrain Soviet behavior—during the Berlin, Suez, and Lebanese crises, for example— and this in itself increased the perceived feasibility of nuclear war.*

* By the early 1960s, American strategic superiority meant that while the United States might have suffered great damage in a nuclear war, this damage would not have been fatal, or even necessarily devastating to a great part of the country. Indeed, in many reasonable scenarios, the U.S. would have suffered relatively little serious damage. In any case, the U.S. would have emerged from a nuclear war with a relatively reliable and rapid ability to recuperate.

The Soviets had a very few long-range bombers, and the United States had a strong and efficient Air Defense Command. Soviet long-range missiles were few and took so long to prepare for launch that in most realistic scenarios, they were likely to fall victim to

But probably the most important contributor to the heightened danger of war in the 1950s was the vulnerability and accident-proneness of both the Soviet and the U.S. strategic forces. Despite the relatively low level of alert at which these forces were maintained during most of this period, the possibility of inadvertent as well as deliberate war was probably substantially greater than it is today; in fact, I am willing to argue that the world was lucky to escape the 1950s and early 1960s without any use of nuclear weapons.

The last two decades have witnessed the development of mechanical safeguards (coded "locks" on nuclear weapons), procedural checks and balances (the "two-man rule," under which no one can be alone with a nuclear weapon), improved tactical warning systems, and the diversification and prelaunch protection of nuclear forces, which probably make accidental nuclear war less likely today than ever in the past. Given their obsession with "control from the center," we can assume the Soviets have taken measures comparable to our own to prevent the unauthorized use of nuclear weapons. Moreover, their interest in avoiding an inadvertent nuclear conflict would be as intense as our own during a crisis. The evidence suggests that U.S. and Soviet leaders have been willing to increase the risk that their nuclear forces will not work well in wartime in order to reduce the risk that they might be used when they should not be. The result is that most analysts today find it difficult (though not impossible) to envision nuclear war by accident.

The task of imagining a nuclear war of any sort, then, is not easy. Nonetheless, some defense experts accept the existence of a "window of vulnerability," a hypothesis that the Soviets have, or soon will have, a significant strategic advantage over the United States that supposedly makes this country particularly vulnerable to certain kinds of Soviet nuclear attacks. These defense analysts and military officers argue that the Soviets may feel compelled to use their nuclear superiority before it erodes (in the late 1980s or early 1990s) when new or modified U.S.

U.S. strategic offensive forces: the Soviets had nothing even remotely resembling an effective emergency launch posture. The few Soviet ballistic missile submarines were quite vulnerable to U.S. naval and air forces, which were more numerous at the time than they are at present. Finally, had any Soviet weapons reached their targets, the United States had, in many respects, far more in the way of survivable recovery capability than it has at present.

These aspects of strategic superiority were not well appreciated at the time. After the early 1960s, the Soviet Union gradually equaled and then exceeded U.S. strategic power, through a combination of their own arms buildup and a leveling off of U.S. force deployments (owing to budgetary limits and unilateral arms-control restraint).

strategic weapon systems are scheduled to be in place. In addition, some who accept the existence of this "window" also note that the Soviet Union is presently facing a number of serious internal and alliance problems (economic, social, political, demographic, and ideological) and might use its military superiority against the United States to distract attention from these troubles.

I agree that a "window of vulnerability" has in fact resulted from the modernization and enlargement of Soviet strategic nuclear forces, that it does indeed represent a "window of danger" for both the U.S. and NATO, and that we should be properly concerned about it. But I do not believe that the Soviets will deliberately turn it into a "window of opportunity" in any dramatic way. Rather, they clearly will exploit their strategic superiority to get whatever political mileage they can, but barring some perverse combination of bad luck and bad management (by either side), or much greater risk taking by the Soviets than now seems likely, the new balance (or imbalance) will probably not result in any watershed changes.

It is interesting to note that studies of the Soviet literature indicate that according to their own concept of strategic balance (what they call the "correlation of forces") the balance began shifting to their advantage in the early 1970s.* In their judgment they have been enjoying the status of this equality and, in some cases, superiority ever since that time. Current U.S. calculations, on the other hand, regard the Soviet strategic advantage as a much more recent and narrow phenomenon.

But even with what they perceive to be a "calculated" advantage, neither the Soviets nor the Americans have a culturally determined and empirically validated tradition of "war by calculation." Unlike the French, Germans, or Japanese before World War II, or the Israelis today, very few U.S. or Soviet military or civilian decision makers will be comfortable with optimistic assessments of their ability to execute a first strike. On the contrary, they are much more likely to accept military analyses that turn out badly than those that predict favorable outcomes. Partly this is a cultural predisposition to distrust war plans, a tendency that reflects the reality of national experiences. While both the Russians and the Americans have achieved major gains through war, neither waged wars that progressed according to prewar plans. Therefore, they believe in "war by miscalculation" rather than "war by cal-

---

* See, for example, Vernon V. Aspaturian, "Soviet Global Power and the Correlation of Forces," *Problems of Communism*, May-June 1980, pp. 1–18.

culation." The prewar Japanese, French, and German experience, on the other hand, reinforced by a much greater degree of national military arrogance and influence, predisposed these nations to accept (at that time) optimistic war plans as a basis for declaring or risking war.

Not many Western or Soviet experts believe that either superpower would deliberately resort to nuclear war to achieve a positive, predetermined ambition, even if projections indicated that the war would very likely be successful.* The uncertainties inherent in the undertaking would be so great that there is little chance that any decision maker in either country would accept the risks (unless absolutely forced to); a deliberate nuclear war would be mostly a war of desperation, a last resort to the least undesirable (if not the only) political-military option.

Even a nuclear war begun in desperation, however, could involve a variety of outbreak scenarios, intrawar alternatives, and potential outcomes. The way in which a nuclear war began and was waged would be extremely important in determining its results. Even a nuclear war can be "won" by one side or the other, and "victory" need not by Pyrrhic. "Win" does not mean unconditional surrender or total devastation for the other side, but merely putting the other side into such a disadvantageous position that it becomes willing to sign a reasonably satisfactory (to the victor) peace treaty rather than continue to wage the war. In this traditional (but not American) view, it is expected and assumed that there will be both bargaining and coercion, as well as destruction and death, and the usual outcome of the war will be not the extremes of total victory or unconditional surrender but some intermediate outcome. The prospect (and nature) of a potential victory might significantly influence the decision to resort to a war of desperation, even if it were not the major determinant.†

A Lesson from the Past

The difficulty of writing "plausible" scenarios for the start of a nuclear war is a cause for some—but not too much—satisfaction. One need only look back to World War I to find a situation where the actual course of events that accompanied the outbreak of war was improbable, im-

---

* Recall my poll of the panel at the National Defense University conference (chapter 2, pp.73–74).

† I discuss a range of possible war outcomes in "Issues of Thermonuclear War Termination," *The Annals*, November 1970, pp. 133–72.

plausible, and sometimes ridiculous (even considering the basic war-proneness of the system of competitive mobilization between opposing European powers). For example:

> During the crisis, the French president was in Russia for a state visit and there were the usual drunken festivities. The Austrians held back an ultimatum because they did not want to present it during a period of intoxicated celebration, when the French President might react in a crazy way.

> While the Austrians were waiting, the German kaiser gave them his famous "blank check" guaranteeing support for any action they took. He then went on vacation! (The kaiser assumed that the Austrian government would not have the courage to abuse this carte blanche; he issued it mainly to avoid being blamed later for any Austrian vacillation during the crisis.)

> The chief of staff of the Russian army received the mobilization order from the tsar. He then thought he would hide so the tsar could not recall the order. (What would be said about a scenario in which the chairman of the Joint Chiefs of Staff made himself scarce so that the president could not countermand an order?)

> The strategy of the German General Staff was to hit the French first, destroy them in six weeks, and then attack the Russians, who would take a longer time to mobilize. At the last minute (August 1), the kaiser, thinking that France and England might still be kept out of a Russo-German war, asked Von Moltke (chief of the General Staff) if the war plans could be changed, and the entire German army shifted to do battle in the East. Von Moltke was aghast and exclaimed, "It cannot be done. We would be left with a disorganized mob, not an army." (Actually, the redeployment probably could have been effected. But what should a political decision maker do when, on the brink of war, his chief of staff informs him that it is impossible to change the existing war plans? Fire him at the moment of supreme national crisis?)

> Similarly, the Russian General Staff lacked the plans and preparations for a partial mobilization directed only against Austria. As a result, the tsar ordered an escalatory general mobilization that threatened both Austria and Germany.

Despite the implausibility of these and other events, World War I actually occurred. The improbability of the outbreak scenario for that conflict (and its all-too-real horrors) highlight the need to consider, study, and plan for the unexpected, and even for the highly unlikely when national security is at stake. In the event of nuclear war, history could again prove to be more cunning and perverse than present studies and analyses would suggest.

Five "Not-Implausible" Scenarios

Hudson Institute has developed a series of basic, canonical scenarios for the outbreak of nuclear war to show how and why deterrence works, and equally important, how and why strategic nuclear forces might be used, depending on the manner in which deterrence failed. The scenarios represent possibilities that, while admittedly remote, are "not implausible."

As mentioned in the Preface, the distinction between "plausible" and "not implausible" is of fundamental importance to the discussion of nuclear war. The term "not implausible" simply means that the scenario cannot be ruled out—there is some significant possibility of its occurring, though that possibility is less than it would be for a "plausible" scenario. The construction puts the burden of proof on an opponent; in order to contradict a not-implausible scenario he must show it to be in fact implausible, which is difficult to do. This is less a debating tactic than a legitimate requirement, for careless or irresponsible arguments or planning in the nuclear era could be disastrous. Even those who believe in nuclear "overkill" should still be interested in how and under what conditions a war might start. Effective remedies first require accurate diagnoses. Any conditions that seem likely to lead to the initiation of a nuclear war are clearly deserving of some thoughtful speculation.

The canonical scenarios sketched below range from surprise attack "out of the blue" (when strategic forces are not on high alert), to war erupting from protracted crises, to attacks occurring after a period of intense competition in arms buildup (i.e., following a "mobilization war"), to conflict resulting from self-fulfilling prophecies (e.g., reciprocal fear of surprise attack). Each scenario raises different issues and presents different challenges to the adequacy of our military planning and our strategic forces.

These five categories of outbreak scenarios provide a reasonable basis for planning U.S. central war posture and mobilization capabilities. They can also be used for educational and polemical purposes and somewhat less effectively as prototypical "triggering scenarios." In international politics, actual triggering scenarios awaken a response which is then a spur to action. The Korean War exemplifies a triggering scenario because it provided the major impetus for a widespread U.S. and allied military mobilization. The response was not so much to counter the North Korean invasion of the South, but it made Congress aware of the need to prepare our forces in the event of a Soviet attack on Western Europe.

The scenarios are presented in order of increasing probability of occurrence (i.e., a U.S.-Soviet mobilization war is a more likely contingency than a Soviet surprise attack against the U.S.). U.S. defense planning should emphasize the last three types of scenarios: protracted crisis, U.S. first strike in the defense of Western Europe, and mobilization war. All of the scenarios are outlined on pp. 234–236 of the Appendix.

SURPRISE NUCLEAR ATTACK

A distinction should be made between a surprise attack that is *largely* unexpected and one that comes *completely* out of the blue. The former is likely to take place during a period of tension that is not so intense that the defender is fully prepared for nuclear war, i.e., during an uneasy confrontation but one where the possibility of escalation has not been considered very seriously. The recent invasion of the Falkland islands by Argentina is a good example of a surprise (albeit nonnuclear) attack. After years of negotiations, the Argentines launched what turned out to be a surprise attack on the Falklands, although the British had been warned that this might happen (but many earlier warnings turned out to be false alarms). In fact, the many warnings neither alleviated Britain's surprise nor increased her preparedness. When the Argentines launched the real attack, they were surprised that the British were actually willing to go to war.

A *total* surprise attack occurs without any immediately apparent or proximate context, although in retrospect one might recognize warning signals that should have been taken more seriously, as in the case of "Barbarossa," Hitler's attack on his ally, the Soviet Union, in 1941.

A distinction can also be made between a deliberate surprise attack and one that is either partially or wholly inadvertent. Under current conditions both kinds are highly unlikely, but more likely than a few years ago. This is due in part to the existence of the "window of vulnerability," and in part to the increased danger of confrontation, which results from what British historian Arnold Toynbee called "withdrawal and return"—in this case, the relative neglect by the United States of its strategic forces from the mid-1960s to the late 1970s and the current "return" via the ongoing strategic force modernization program (which was initiated during the Carter administration and largely strengthened by the Reagan administration).

A deliberate Soviet attack out of the blue would probably be a constrained or "environmental" counterforce attack.* It would probably not include urban-industrial centers in a major way (even though popular images of a nuclear war consider cities to be the primary targets for an attack). If the Soviets started a war, they would be very conscious of the enormous risk they were taking and would make overt efforts to minimize the risk of damage to themselves from a possible U.S. retaliatory strike. In a constrained counterforce attack, the Soviets would intentionally impose certain limitations on the military effectiveness of their attack in order to minimize civilian and industrial damage to the U.S., thus minimizing the threat to Soviet cities and civilians from U.S. retaliation. U.S. decision makers, contemplating a retaliatory attack (with damaged and reduced strategic forces), would face the possibility of an unconstrained, full-scale countercity response in the Soviets' second strike.

Constrained attacks pose many grave problems to both attacker and attacked—primarily, the difficulty of recognizing the nature and extent of the constraint. The United States would find the distinction between a constrained attack and an all-out one especially difficult if, as part of their initial strike, the Soviets included a "decapitating attack" on the U.S. command-and-control system. In this type of deliberate surprise attack, however, the Soviets might be able to provide the U.S. with verifiable information on the "limited" nature of their attack as soon as it was under way. Indeed, it would be in their own best interest to do so. Since the Soviets would have little incentive to lie (and the U.S.

* An example of an environmental counterforce attack would be the detonation of nuclear weapons at high altitudes over an enemy's homeland, with the aim of creating electromagnetic pulses sufficient to disrupt and disable military communications, thereby making a retaliatory attack difficult to execute.

could probably check up on them relatively easily), the U.S. would likely accept the Soviet statement—at least tentatively.*

The Soviets might also incorporate a hedge, or a "two-layer strategy" into their first strike. The first layer would involve a constrained attack on the command-and-control network, designed to prevent a U.S. nuclear reprisal. If it were successful, the Soviets could claim a victory. If it failed, the second layer would be an announcement that the partial effects of the attack were because of intentional constraints, and the U.S. should therefore exercise similar caution when retaliating. But the Soviets could not count on U.S. restraint. If many American cities were severely damaged, and millions of American lives were lost in the initial onslaught, an American president would be unlikely to respond with restraint, even if it meant risking the lives of another 100 million Americans. On the other hand, if there were no attacks at all on U.S. cities, or if these were in fact quite limited, it is difficult to imagine a president launching a large countercity retaliation—particularly if the Soviets were demanding less than unconditional surrender by the U.S. (although they might demand something close to that from the West Europeans).

Of course, even with a constrained U.S. nuclear retaliation, a war could become unlimited at any time. The choice of strategy and tactics is not solely an initial decision, and the point at which a war becomes unrestrained can make a major difference in its consequences. Soviet military analysts have no doubt noted that a nuclear war of reciprocal countercity strikes would do less damage to the Soviet urban-industrial complex if it occurred relatively late in the war rather than during the first or second exchange. Late escalation to urban-industrial attacks would give the Soviets more time to disperse and shelter a larger portion of their population, and even more important, the U.S. would then be attacking with damaged and poorly coordinated forces.

In what is perhaps the most common scenario for a "restrained" surprise attack, the Soviets eliminate nearly all U.S. land-based mis-

---

* If the Soviets did lie, it would probably be in an attempt to delay U.S. retaliatory attacks against their cities. That is, the Soviets would launch a more or less comprehensive attack on U.S. urban-industrial centers as well as military installations. By providing false information indicating that the attack was a constrained counterforce strike, the Soviets would try to confuse U.S. efforts to piece together an accurate attack assessment. If the Soviet claim were believed, the U.S. would probably retaliate with its ragged surviving strategic forces against only those Soviet military forces held in reserve. Thus, if this lying tactic worked, the Soviets would have hit the U.S. with a relatively unrestrained attack, while suffering a much lesser and more constrained U.S. attack in return.

siles, part of the long-range bomber force, and a portion of the fleet of ballistic missile submarines (mostly those in port), while simultaneously trying to decapitate the U.S. response system through strikes on command-and-control facilities and equipment. Targets located near large urban population centers are attacked with special munitions (e.g., nuclear warheads with carefully tailored yields or effects) or tactics that cause little collateral damage. In one variant of this scenario, the Soviet attack is immediately followed by an ultimatum to surrender, and the U.S. surrenders, or at least does not retaliate. In my view, either response is most unlikely: the U.S. would still have a significant, albeit damaged, retaliatory force of bombers and submarines with which it would almost certainly strike back.

The question, then, is "which Soviet targets should the U.S. strike back at?" If it attacks Soviet cities (where most of the population may have been protected in advance of the U.S. retaliation), the U.S. will almost certainly suffer great damage to its own cities (and people) in a counterblow. But if the Soviets had demanded unconditional surrender, it is unlikely that the U.S. would pursue a limited retaliation that avoided Soviet cities. The launching of an all-out strike against the Soviet Union would be much more likely.

Retaliation against Soviet strategic offensive forces would be difficult for the United States to carry out, however, even if the Soviets limited their initial attack to a constrained first strike against military targets. For example, if U.S. command-and-control assets were destroyed or disabled, it would be very hard to know which Soviet missile silos were not empty. The U.S. would, however, still be able to go after what it considered to be weak links or "leverage targets" in the Soviet force (e.g., some portions of *their* command-and-control system), even if the retaliation were of reduced or uncertain effectiveness.

The point is that a U.S. failure to retaliate would be psychologically and politically unacceptable. Inaction would make it very difficult for the U.S. to bargain and negotiate credibly to end the conflict on favorable terms. The "capitulation scenario" is therefore inherently flawed, and its widespread use has cast doubt on, rather than supported, the validity of the window-of-vulnerability concept.

This issue of bargaining versus surrender raises basic questions about the scenario for a *constrained* surprise attack: If the Soviet Union were really seeking unconditional surrender by the United States (i.e., the elimination of the U.S. as a threat to Soviet security or to a future worldwide socialist commonwealth), why would it exercise restraint in its first strike? Alternatively, if it were not seeking unconditional surrender, why would it undertake the massive risks and uncertainties of nuclear war?

These questions are legitimate, but not conclusive. They do not make the scenario completely implausible. One answer to why the Soviets exercised restraint at all might postulate that even though the Soviets were seeking stark unconditional surrender (if not immediately, then eventually), they believed it would be easier to achieve through a constrained attack than through a strike that left the U.S. no options except immediate surrender (highly unlikely) or a strike against Soviet cities (highly injurious). For the Soviets, a constrained attack would buy time to get in another strike which could be very helpful in achieving their ultimate goal.

The answer to the second question—why risk war at all for less than unconditional surrender—is easier and more persuasive: the Soviets might be willing to risk war if some immediate crisis threatened their short-term future, and only incidentally because they sought ultimate world domination or greater long-term security.

If the Soviets' only objective were the total destruction of the United States, they could launch an environmental attack against people and property designed to take advantage of both expected and unexpected (or at least uncertain) primary and secondary nuclear-weapons effects (long-term radiation, widespread fires, tidal waves, extensive blankets of nuclear blast, and so on). An environmental attack is in some ways quite consistent with our understanding of Soviet motivations and behavior, but because it is much more risky than a constrained counterforce attack (i.e., it probably could not physically preclude a devastating second strike by well-protected U.S. strategic forces), we consider it less likely.

EARLY ERUPTION TO NUCLEAR WAR FROM AN INTENSE CRISIS

As noted, this seems a much more plausible scenario for the initiation of nuclear war than a surprise attack. But even where the crisis itself is not implausible, it is still quite difficult to develop a chain of subsequent events leading to nuclear war. Nonetheless, four situations seem "not implausible" candidates for a direct U.S.–Soviet confrontation that could conceivably escalate to nuclear war. These are: (1) an uprising in

Eastern Europe with NATO somehow becoming involved in the crisis; (2) Soviet military intervention in the Persian Gulf region followed by a U.S. counterintervention; (3) a Sino-Soviet war with the U.S. allying itself with China; and (4) some other East Asian, third-party conflict. Of the four crises, the first two seem to have the greatest plausibility.

Most scenarios that envision a nuclear eruption from a crisis in Eastern Europe assume a limited but intense confrontation that threatens to get out of control and spread chaos and rebellion to all of Eastern Europe, or even to the Soviet Union itself. In a desperate attempt to prevent this (and the inevitable loss of political control by the Soviet regime), the Soviet Union would attack Western Europe or the United States. In the specific case of an East German uprising, escalation to war might begin with the frontier between East and West Germany opening up, and sporadic fighting breaking out between West German (NATO) units and East German (Warsaw Pact) troops.*

Potential U.S.-Soviet conflicts in the Persian Gulf offer somewhat less persuasive scenarios for the outbreak of nuclear war. American planners sometimes speculate that the U.S. might initiate the use of nuclear weapons in a similar desperate attempt to stave off defeat (i.e., to prevent a Soviet invasion force from overwhelming U.S. intervention forces defending the Gulf oil fields). Another desperation-based U.S. option is known as "horizontal escalation." This would involve starting a new crisis or battle elsewhere (e.g., an attack on Cuba), in order to divert Soviet attention and resources away from the immediate theater of war, or to inflict comparable (if not offsetting) damage to Soviet worldwide security interests. But neither the straight nuclear option nor a serious horizontal escalation seems very plausible, even if the alternative is defeat.†

The difficulty of writing plausible outbreak scenarios notwithstanding, crises can get out of control and can force one side or the other to become desperate—perhaps to the point of deciding that striking first is safer than not striking first (if only to preempt the other side). This would be a negative, "last-resort" calculation. While the calculation might be influenced by the hope of achieving certain positive goals (e.g., the Soviet dream of creating a worldwide socialist state in the aftermath

---

* See "An East German Uprising Scenario," in Frank E. Armbruster, *A Study of Some Issues Which May Influence the Question of a Role for Free-Fall Weapons* (Croton-on-Hudson, N.Y.: Hudson Institute, 1981), pp. A–4 through A–58.

† Chapter 7 includes a discussion of how a U.S. strategic mobilization would be preferable in this case to both nuclear use and horizontal escalation (see p. 159).

of a final confrontation with capitalism), it is doubtful that such ambitions would by themselves trigger a nuclear war.

Alternatively, one side might consider that a crisis was not desperate enough to *compel* drastic action, but still presented a good chance to eliminate the other side as a source of future trouble. In this case, the conflict would be a preventive nuclear war. Under most circumstances, "preventive nuclear war" is almost a contradiction in terms. Bismarck's remark that preventive war is like committing suicide for fear of death aptly describes the unattractiveness of this alternative in the context of a long-term competition between two nuclear-armed superpowers. Nonetheless, one can imagine extraordinarily intense U.S.-Soviet confrontations in which the motivations of preventive war might help tip the scales in favor of the resort to nuclear use. Again, it should be stressed that this justification would most likely be only part of a last-ditch rationale.

The critical element in a decision by either side to initiate nuclear war would be the judgment that attacking was less dangerous than not attacking. This judgment might be faulty but it need not necessarily be irrational. The only time this might be an almost instinctive, nonrational decision would be in a crisis where the Soviet regime's near-paranoid fear of losing internal political control superseded any other consideration.

However, this same Soviet paranoia about maintaining political control could also serve as a check on the aggressive tendencies of the Soviet state. For example, even in a serious confrontation, the Soviet regime might prefer to lessen its pressure on external enemies in order to avoid a catastrophic war that might result in counterrevolution. Conversely, as long as they felt they could retain political control and ultimately triumph, Soviet leaders might be willing to risk nuclear war, despite the massive physical damage such a conflict would cause. Indeed, in the Napoleonic War and World War II, Russia suffered terrible destruction, but after both experiences rapidly emerged stronger than before. As a result of World War I, however, the Tsarist government was defeated and overthrown, and the Bolsheviks ultimately seized power. To the extent, then, that the United States and its NATO allies can threaten the Soviets with a loss of political control as a result of a nuclear confrontation, the West might increase its capability for deterrence quite markedly.

Let us make the reasonable assumption that the Soviets fear the political consequences of an ongoing crisis more than they fear the physical consequences of any retaliation for an attack on the United States. Further, assume that the Soviets would prefer to resolve the crisis through negotiation rather than through war. Accordingly, the most likely action by the Soviets would be to order a large-scale evacuation of their major urban centers, a move calculated to give them crucial leverage in their bargaining with the United States. With most of their people removed to places of relative safety, the Soviet leaders could then threaten to launch a partially disarming counterforce first strike. The U.S. would have few appropriate retaliatory options, since its own main counterforce weapons (the land-based ICBMs) and much of its command-and-control system for wielding those weapons effectively would be vulnerable to such a strike. Any retaliation against Soviet cities would destroy mostly empty buildings. (Of course the Soviets would value even empty cities; these would retain much administrative, economic, historical, and aesthetic value. But since most of the cities eventually could be rebuilt, they might be risked if necessary.)

It is sometimes argued that the Soviet Union would not evacuate its urban population in order to avoid warning the U.S. that a nuclear attack was imminent. Certainly a Soviet evacuation would be treated as one kind of strategic warning; U.S. strategic forces would be placed at a higher level of alert to reduce their vulnerability to Soviet attack. But the Soviet leadership might accept this disadvantage because of the offsetting advantages: most of its population would be protected, and it would be in a superior bargaining position. (In fact, to a certain degree, putting the U.S. and its allies on notice would itself be advantageous; as noted, evacuation would be a means whereby the Soviets could coerce the Western powers into a resolution of the crisis short of war.)

A Soviet evacuation would also pose the risk of a preemptive strike by the U.S., but given the current balance of nuclear forces, this possibility seems very remote. While U.S. counterforce operations might be able to reduce the threat posed by Soviet offensive forces, those strikes could not effectively disarm the Soviet Union. A U.S. president would therefore find it very difficult to justify the actual use of nuclear weapons prior to a direct attack on the U.S. or its allies (especially if the no-first-use concept gains more adherents). But a large-scale Soviet evacuation could prompt a U.S. counterevacuation—a most appropriate, if dangerous, response. The United States currently has only minimal preparations for civil defense, but Hudson Institute studies suggest that this handicap is not overwhelming.* An American evacuation effort, even if mostly improvised, could still be remarkably effective in rapidly reducing the number of expected casualties after a nuclear attack (although less effective than one based on a stronger peacetime civil defense program).

With an evacuation ploy, therefore, the Soviets would have to consider the possibility of a U.S. preemptive strike or counterevacuation, but in many potential crises the Soviet Union would probably not be deterred if these were the only counterthreats. There are of course others, and again, some analysts feel that among the most serious Soviet worries would be that a massive evacuation could precipitate a loss of political control (through the disorder and anxiety that might accompany the relocation of millions of Soviet citizens during a nuclear crisis). However, it seems unlikely that any significant, organized domestic challenges to Soviet power would result; in fact, if a short-term evacuation turned out reasonably well, the actions of the Soviet government would appear both competent and justified to the Soviet people.

Despite this assessment, the overall dynamics and consequences of a large-scale Soviet crisis evacuation—its economic impact, its social and political strains, and perhaps most important, its political aftereffects whether or not a war occurs—are significant enough to merit more extensive analysis than Western observers have devoted to them.

This is especially true since the chances of inadvertent war may increase greatly if either side evacuates, and perhaps increase even more if both sides do. And even if neither side evacuates with the intention to strike, the fear of war, as well as the military and psychological preparedness for war, could escalate dramatically. The fear (in the United States and the Soviet Union, in Europe, and probably throughout the rest of the world) could lead to enormous pressure, particularly on the United States, to compromise or capitulate. Clearly, there would be many problems on the way to de-escalation by either or both sides. The tensions, actions, reactions, and misunderstandings could at any point lead to inadvertent nuclear war.

The dangers of an accidental war would be greatest if one or both sides moved to a "launch-on-warning" † posture during an intense crisis. But even without the adoption of this hair-trigger tactic, the possibility of inadvertent escalation is especially great during any crisis

---

* See, for example, William M. Brown and Doris Yokelson, *Postattack Recovery Strategies* (Croton-on-Hudson, N.Y.: Hudson Institute, 1980).

† See footnote on p. 50.

where the combat readiness of U.S. and Soviet nuclear forces is greatly increased. In theory, the increased danger might foster greater caution, but caution is not always psychologically consistent with the tension, pressures, and other demands of a high alert. Incidents that might create only minor problems under normal conditions could prove extremely volatile during a crisis. The Soviets are very aware of the heightened risk of irresponsible or unauthorized behavior during a high alert and have consequently never ordered one for their strategic forces (even during the Cuban missile crisis).

The evolution of a war-prone crisis and the subsequent efforts by both sides to find bargaining leverage could easily lead to various "demonstration" or "symbolic" nuclear attacks designed either to warn the other side to back down or to achieve a specific, limited military goal. It is sometimes argued that "coercive diplomacy" employing physically "harmless" demonstration attacks (say, a low-kiloton weapon burst high over the opponent's capital city) would be counterproductive, indicating an overly theoretical and cautious attitude toward central war rather than emphasizing true determination and resolve. But it could also be a very effective signal of the gravity of the crisis and one's commitment to ultimate success. If, for example, West German troops intervened without government authorization in an East German uprising, a small Soviet nuclear detonation a few hundred thousand feet above Bonn could send a very clear message (without doing any physical damage) that the provision of official West German assistance would be a serious error. It might also force a halt to the East German revolt.

A militarily useful demonstration attack could take a number of forms. For example, the Soviets will probably soon be able to find and destroy at least a few American ballistic missile submarines at sea. By eliminating even a small part of the U.S. retaliatory force during a crisis, the Soviet Union would have gained an important bargaining advantage by raising doubts about the invulnerability of the rest of the U.S. sea-based deterrent. Or the Soviets could militarily "demonstrate" the seriousness of their intentions by attempting to use the electromagnetic pulses (EMP) produced by nuclear explosions to knock out U.S. electrical equipment (e.g., military communications components). They could do this by detonating, say, five multimegaton weapons at a high altitude over the United States. The effects of such an attack cannot be predicted with great confidence; it might do little damage, or it might wipe out all but the most protected circuits. If the attack were overwhelmingly effective, the Soviets could follow up with an ultimatum or with another attack; if ineffective, they could present it as a pure exemplary attack, intended solely to discipline U.S. crisis behavior but not to cause permanent or retaliation-inducing damage to people or property.

There is a broad range of limited nuclear-attack options available. The "safest" demonstration strikes are probably attacks at sea or in space. But any use of nuclear weapons clearly increases the risk of escalation to higher levels of conflict.

U.S. FIRST STRIKE TO DEFEND WESTERN EUROPE

This is the "classic" scenario for the outbreak of nuclear war between the U.S. and the Soviet Union. It emphasizes the "extended deterrence" role of U.S. strategic nuclear forces and raises the issue of the capabilities required to fulfill this role. At present, the U.S. guarantees its NATO allies that it will use nuclear weapons to defend Western Europe against any Soviet aggression, including conventional attack: that pledge essentially constitutes the "extended deterrent" we have provided for our allies throughout the postwar period.

In the late 1940s and early 1950s, the fundamental mission of the Strategic Air Command was, in fact, less to defend the continental United States than to defend Western Europe from a Soviet attack. But the credibility of the U.S. extended deterrent has steadily eroded with the development of a Soviet capability to launch large-scale nuclear strikes against the United States, and especially with the Soviet attainment of strategic nuclear parity with the U.S. and improvements in Soviet theater nuclear forces targeted against Western Europe. Nevertheless, the nuclear guarantee of the defense of Western Europe remains a basic (although increasingly questionable) premise of U.S. and NATO military policy.

If the U.S. is to continue to extend (Type II) deterrence to NATO, it must to some extent restore the credibility of its threat of a strategic first strike in response to Soviet military aggression against Western Europe. The Soviets must be persuaded that the U.S. not only possesses appropriate nuclear forces but also the will to use them in response to an attack on its NATO allies.

I have argued that this calls for a not incredible counterforce first-strike capability. Each term in this (somewhat awkward) label is important. The not incredible means that it must be sufficiently credible that the United States can be provoked into launching what is a first strike as far as the strategic forces are concerned, but in a second (or later) strike as far as major attacks (e.g., the conventional invasion of NATO-Europe) are concerned. The credibility is measured by the actual deterrent effect on the Soviets, by the assurance it provides our allies, and by the assurance of our diplomats in and out of conferences. The periphrasis "not incredible" implies that even a relatively low level of credibility will suffice for all of the above. The announced use of counterforce operations (combined with planning and tactics for controlled central war and intrawar deterrence) is one of the ways of increasing both the credibility and actual utility of the posture.

Given the present (and likely future) U.S.-Soviet strategic balance, a complete U.S. disarming-strike capability would be very difficult (if not impossible) to achieve or maintain. However, in most circumstances the United States will still have a capability for significant "force-reduction attacks" against Soviet strategic offensive power. A not incredible counterforce first strike is one such attack.

A force-reduction attack is one that puts out of commission an important part of the enemy's forces (e.g., ICBMs in silos, bombers at air bases, and submarines in port), thereby changing the balance of forces in a major way. A force-reduction attack does not eliminate the enemy's capability for a devastating second strike against the attacker's cities—by definition it is a partially disarming attack. But the attacker can deter this kind of retaliation by threats of equal or greater counterretaliation, and meanwhile prepare to carry out further force-reduction attacks, attempt to reduce his own vulnerability to retaliatory attacks, and offer to negotiate an end to the hostilities.

If, in response to a Soviet attack on Western Europe, the United States hit the Soviet Union with a not incredible counterforce first strike (that tipped the military balance against the Soviets and destroyed their chances of winning) and issued a blackmail threat to increase the damage if they retaliated against U.S. cities, the Soviets would tend to negotiate terms (e.g., a return to the status quo ante) rather than continue the war.

No U.S. president could credibly threaten to "punish" the Soviets for an attack on Western Europe if the Soviets could then retaliate with nuclear strikes that deliberately or (as a result of a counterforce attack) incidentally killed the bulk of the American population. The U.S. must, therefore, also have adequate civil defense capabilities, especially the ability to evacuate cities during an intense crisis. Physical destruction of economic assets and infrastructure can be repaired as long as the manpower and the technical skills (the country's greatest source of wealth) still exist and can function.* In the future, the prospects for U.S. recuperation from a nuclear attack could be further enhanced through the deployment of a ballistic missile defense system. Improved strategic defenses, then, would strengthen the credibility of U.S. first-strike threats, and thus the deterrence of extreme Soviet provocations.

Some provocations may initially elicit only a verbal warning; a not incredible counterforce first-strike threat need not be carried out instantaneously. One can imagine a Soviet conventional attack on Western Europe that in the first instance produced only a U.S. ultimatum that the Soviets withdraw. At that point, having sufficiently credible military forces, strategic defenses, and political resolve would permit the U.S. to bargain for an acceptable settlement without resorting to central war. The severity and character of a Soviet provocation in Europe would be almost as decisive as U.S. preparations in determining the nature—and credibility—of a U.S. nuclear reprisal.

In chapter 10, I propose that the most appropriate NATO response to current questions about the U.S. nuclear guarantee would be to build up the conventional forces of the Alliance, to pledge no-first-use of nuclear weapons, and to improve the theater nuclear forces in Europe. These measures, coupled with the strengthening of the U.S. strategic deterrent, would increase the effectiveness of a conventional NATO response to conventional aggression, and the credibility of a nuclear response to nuclear aggression. The role of a U.S. not incredible counterforce first-strike threat would then be restricted to the deterrence of *Soviet* first use of nuclear weapons against the NATO allies.

A no-first-use declaration unaccompanied by nuclear rearmament would further strain the credibility of the U.S. nuclear guarantee, as well as diminish the cohesion of the Alliance. Without a credible U.S. nuclear deterrent, any conventional European rearmament would lose its raison d'etre. If the United States could not reliably protect Western Europe from a Soviet *nuclear* attack, why should the Europeans bother to make the considerable sacrifices needed to provide an adequate *conventional* defense? Thus, the assurance that the United States would deter Soviet nuclear aggression would have to underlie any conventional European rearmament.

In some ways, the United States can eat its cake and have it too by declaring no first use. A no-first-use pledge genuinely would put U.S. and NATO nuclear-weapons policies on a sounder moral and political

---

* See chapter 8 for further discussion of civil defense and postattack recovery.

footing. At the same time, the effectiveness of Alliance military preparations need not be compromised by the renunciation of first nuclear
use. Prudent Soviet military planners would have to discount the
pledge, especially if it were accompanied by a bolstering of U.S. nuclear
war-fighting capabilities. Consequently, U.S. strategic forces would retain an important degree of implicit credibility as a deterrent to Soviet
conventional aggression. Equally prudent European defense officials
and parliamentarians *would* believe the pledge, and argue that, with
U.S. strategic forces no longer substituting for NATO's weaknesses in
conventional defense, European nonnuclear forces would need to be
significantly improved. Thus, in the real world there would not be the
sharp contradictions that seem to exist among no-first-use, first-strike
threats for extended deterrence, and a West European commitment to
strengthened conventional forces for deterring Soviet conventional
attack.

ESCALATION TO NUCLEAR WAR FROM A PROTRACTED CRISIS

This scenario evokes the possibility of one of the aforementioned U.S.-
Soviet crises, or a period of confrontation, or even a limited but escalation-prone conflict, extending over a long time. A prolonged crisis
raises issues and problems different from the other outbreak scenarios,
primarily because neither U.S. nor Soviet strategic forces are designed
to withstand the special strains of an extended alert. The common (and
probably faulty) conception of a nuclear war starting soon after a crisis
begins, and lasting minutes (or at most, hours) has dominated the design
of nuclear forces and the planning for their use. Relatively little thought
has been given to the probable erosion of strategic force capabilities
over time—the tremendous physical and mental toll that would be imposed by an extended high alert. Yet these factors would become significant in a protracted crisis, and would influence the options available
and, ultimately, the resolution of the crisis.

Compared with peacetime, the chance of inadvertent war would be
much greater during a protracted crisis in which the war-readiness of
nuclear forces had been increased. In fact, in a long and tense confrontation, inadvertent war might be the most likely scenario for nuclear
conflict. Both sides would be under tremendous pressure to either attack or reach a settlement. In addition, fluctuations in the level of tension, that are bound to occur during a lengthy crisis, would be extremely
destabilizing. Fatigue (of both men and equipment) could dull reactions
and produce serious misjudgments and malfunctions. The uncertainties

and stresses of shifting tensions could yield both under and overreactions, complacency, or supersensitivity—with possibly disastrous consequences. Either side could exploit the strains, for example, by using a "lull" in the tension to suggest that a resolution of the crisis was likely, and then attacking while the other side's guard was down.

Some form of relatively limited conflict—either conventional or nuclear engagements in Western Europe, at sea, or in space—could occur in a protracted crisis on either a sporadic or a sustained basis. The action could be fairly large scale, in which case the chance of an eruption into a central nuclear war would be high; or it could involve more restrained acts of attrition war, in which case each side would try to degrade or destroy the other's military capabilities without provoking a full-scale reaction. Selective attacks might be conducted to diminish the effectiveness of the enemy's command-and-control network. "Spoofing" tactics (i.e., moves that feigned an attack) might be employed to dull the responsiveness of the network, making the enemy more vulnerable to what I referred to above as a "largely unexpected" surprise attack. All of these ploys are quite dangerous, but then almost any military action in a protracted crisis carries the risk of escalation (though some carry greater risks than others).

ESCALATION TO NUCLEAR WAR FROM A "MOBILIZATION WAR"

As described in the next chapter, mobilization involves a rapid increase in force deployments and the large-scale conversion of a nation's economic, technological, and organizational resources into military power. Mobilization is usually initiated in response to a serious crisis or conflict. A "war" of competitive mobilization avoids immediate high-intensity violence while the opposing nations use their ongoing military buildups for bargaining purposes during the crisis. If the mobilization war does not result in a resolution of the crisis without direct conflict, then, as a consequence of mobilization, one or both antagonists will be better prepared to fight when real war does begin.

Although the Hudson Institute has done a number of studies of how a mobilization war might be conducted,* we find that a great deal must

---

* See, for example, Herman Kahn and William Schneider, Jr., *The Technological Requirements of Mobilization Warfare*, 1975; William Schneider, Jr., *Strategic Mobilization as an Element of Defense Policy Planning*, 1977; and Paul Bracken, *Mobilization in the Nuclear Age*, 1978.

still be learned about the dynamics and consequences of such a political-military competition. Like other arms buildups, mobilization need not erupt into actual fighting, but will almost certainly lead to an alteration in the military balance. A mobilization activated today in response to a serious U.S.-Soviet crisis would probably place the United States in a dominant strategic position, ultimately forcing the Soviets to back down from whatever act of aggression provoked the mobilization. Alternatively, the mobilization could put the U.S. in a much better position to respond to any outbreak of war, should the Soviets choose to launch a preemptive first strike or try to exploit the initial crisis further.

The effort to increase one's own strength relative to that of the enemy during a mobilization is much the same as in a direct conflict. In either case, the U.S. would seek to maximize its present and future "threat potential" against Soviet military forces and political values, while minimizing the Soviet threat to the West. Of course, if the Soviets felt their military situation vis-a-vis the United States were rapidly deteriorating, they could step up their own mobilization, settle the conflict quickly, or strike first out of desperation.

If a U.S.-Soviet central nuclear war broke out roughly one year after a mobilization war began, I would judge that the United States by then could have acquired a relatively effective nuclear war-fighting posture. In particular, there would have been time to make substantial improvements in command and control, provide most of the American population with fairly good civil defense protection, and increase the readiness and usability of the strategic forces (despite the problems posed by a protracted crisis). Within two years the U.S. could have a greatly improved war-fighting posture, based on a sizable expansion of existing strategic forces. After three years of mobilization war, the U.S. might be able to field genuinely new systems (e.g., a multilayered missile defense system).

Problems might arise if a protracted crisis coincided with a mobilization war. In this scenario the country would be maintaining a high state of alert (and possibly carrying out crisis evacuation and other emergency measures to protect the population), while at the same time gearing up for a major mobilization effort (i.e., very high levels of defense production). The two activities would interfere with each other to a degree, but it should be possible to work out reasonable trade-offs and develop a relatively efficient approach for dealing with both (although additional study of the issues is clearly required).

A Postscript on Soviet Prudence and Caution

As noted, the various judgments incorporated into the five canonical scenarios assume that the Soviet leadership is essentially prudent and cautious. For example, despite the significant advantages the Soviets might reap from a surprise attack (out of the blue) against U.S. strategic forces, I have assigned this danger a low order of probability, largely because the Soviets are not that reckless. If, for some reason, Soviet leaders turn out to be less circumspect than we have assumed, many of the theories of nuclear deterrence and war fighting would have to be modified (in some cases, drastically). In particular, as circumstances change, Soviet concepts of prudence and caution may change, and in ways we may not realize. Or worse, Yuri Andropov, his successor, or even the entire Soviet ruling elite may not be as prudent and cautious as we think.

Thus, the characteristics and intentions of the Soviet leadership are fundamental and important issues. There is a wide spectrum of views about the motivations of the Soviets. Views range from conceptions of the Soviets as benign to rational to opportunistic to messianic to demonic. To a considerable extent, some aspects of Soviet intentions are unknowable. To the degree that Soviet behavior can be predicted, it is useful to think about what might happen if a different kind of leadership came to power in the Soviet Union. One way to do this is to employ historical characters—Alexander the Great, Napoleon, Hitler, Frederick the Great, and Wilhelm II—as models for potential (however improbable) Soviet leaders. These models illustrate possibilities that have been ignored or underemphasized in the five canonical scenarios.*

ALEXANDER THE GREAT

He was raised to implement the ancient equivalent of world conquest. Alexander took over a well-functioning enterprise, Macedonia, built up by his father, an ambitious and competent man who created a first-class army, conquered Greece with little bloodshed, and was preparing to move against the Persian Empire when he was assassinated. Alexander, who was carefully trained to rule (by Aristotle, among others), executed the plan brilliantly, being as ruthless as necessary but preferring to

* I am indebted to my colleague B. Bruce-Briggs for his help in developing the models.

pacify conquered peoples by key concessions to local sensibilities. Alexander very nearly conquered all the world then known to his contemporaries; he failed to complete the job only because his troops mutinied and he died prematurely.

Marxism-Leninism has a similar vision (and theory) of "world conquest"—the emergence of a world socialist commonwealth. Today, with nuclear weapons and modern communications, the establishment and policing of a true world imperium is a practical possibility. But what are the chances of an "Alexander the Great" taking control of the Soviet Union? Practically nil, we would argue. The Soviet system is controlled by a self-perpetuating gerontocracy that precludes a young, skilled, and ambitious warrior from coming to power. Soviet military forces lack the kind of superiority in morale, competence, and arms that the Macedonian army possessed relative to its adversaries. The Soviets certainly lack the sensitivity in handling subject populations that was so important to Alexander's success. Finally, the dynamic personality associated with Alexander the Great, and the temperament to take risks, seem to be very much suppressed and rejected in Soviet society, which instead encourages personality traits such as conformity and caution. While the Soviet system does recognize and promote those who seize opportunities, or help create them, it shows no signs of being receptive to someone who would initiate a high-risk calculated war, or someone who would be motivated almost entirely by an overwhelming personal ambition to succeed in world conquest.

NAPOLEON

Napoleon took over an ongoing revolution and turned it into a personal dictatorship. In war he was the consummate strategist and tactician. He was utterly ruthless but not cruel—he took no satisfaction from seeing suffering, but was indifferent to it so long as it served his purposes. He was feared, but not respected.

Employing the ideological trappings of the French Revolution, Napoleon's armies were initially welcomed as liberators, but his policy of supporting his wars by looting his conquests turned Napoleon the liberator into a tyrant. His excessive ambitions, which caused him to make enemies out of potential supporters, in the end brought him down.

Can we have a Soviet Napoleon? It is rather late in history for that possibility. General Bonaparte tapped the ongoing dynamism of the French Revolution and was able to attract hundreds of ambitious and talented young officers and officials. The Bolsheviks were all too aware of the Napoleonic model. Their fear that Red Army chief Leon Trotsky might be their Bonaparte was a major reason for the triumph of his rival, Joseph Stalin. The current fear of another Stalin is much greater than the fear of a Soviet Napoleon, and makes the emergence of a Napoleon type even less likely. Thus, although a Soviet Napoleon is more plausible than a Soviet Alexander, the structure of the Soviet system virtually precludes the appearance of any individual (or group) with Napoleonic talent and ambitions. It is true, however, that the Soviets emulate Bonapartian ruthlessness in their domestic and international affairs.

HITLER

This is the favorite modern model of an aggressor. In fact, the recollection of Hitler's aggressions and the failure of British and French attempts to appease them was a "lesson of the past" that informed much of Western military policy at the beginning of the cold war. However valid Hitler may have been as a metaphor for Stalin, he is far less valid as a prototype for the present Soviet leadership.

Like Alexander and Napoleon, Hitler was able to attract able young men to his banner. Nonetheless, Nazism was a movement he personally created, with an action program that he believed only he could carry off. (In a certain sense, he was right.) His ego compelled him to move rapidly to achieve the domination of Europe while he was still active.

Hitler was a master of the use of terror in his political and military tactics. Unlike Alexander, he made little attempt to accommodate most of the peoples he conquered, though the French, Dutch, and Scandinavians received better treatment than the Slavs. And unlike Napoleon, who was ruthless but not cruel, Hitler was cruel rather than ruthless. His policy was too much influenced by personal sentiment: for example, his admiration of the English led him to postpone escalation to total war against Great Britain; his contempt for the French ruined a valuable collaboration with French reactionaries; his racism prevented him from exploiting his Ukranian collaborators better; and his idealization of German domestic life contributed to his reluctance to fully mobilize the Third Reich for war.

Hitler was also odd for a twentieth-century leader in that he glorified war—a characteristic markedly lacking in the political doctrine of Marxism-Leninism. Unlike his Soviet counterparts, Hitler began his political life in a liberal democratic polity, and thus understood (and despised) bourgeois values and bourgeois politicians. This helped him to become a master of blackmail. He could prey upon bourgeois politicians' aversion to violence and their sense of fairness. He ranted and raved about the horrors of war (which would be suffered by nations standing in his way) and the rights of Germans in order to extract major concessions from his adversaries. Eventually Hitler's demands became intolerable for most of his opponents.

A Soviet—or any other—Hitler seems extremely implausible because men with such capacity for evil are fortunately rare. The oligarchical character of the Soviet leadership in particular, however, would again seem to create insuperable barriers to the emergence of such an individual. It is impossible to imagine a young Hitler working his way up through any of the Soviet bureaucracies. (Even after four years of service, the Imperial German Army would not trust the real Hitler to command so much as a squad.) But even more important may be the fact that the Soviet system is not receptive to the kind of fascism that Hitler represented. It has been argued that without defeat in World War I, the hyperinflation of the 1920s followed by the Great Depression, and the weaknesses of the Weimar Republic, Germany would not have been receptive to Hitler either. It was the culmination of these events that led the Germans to accept Hitler as a leader and convinced them to take the risks that he represented. A confluence of similarly disastrous events is unlikely to occur in the Soviet Union.

FREDERICK THE GREAT

Here is another great captain who upset the stability of Europe. Neither ruthless nor cruel, Frederick was a complete cynic, with nothing but contempt for humanity; expecting little of his fellowman, he was never disappointed or angry. He was the master of calculated aggression for limited objectives. He favored tactical surprise attacks in protracted crises. For example, when he learned of a coalition of his enemies, Frederick launched a classic preemptive strike against one of them, Saxony, and proceeded to take over its resources and its army for use in the ensuing Seven Years' War.

With regard to models for the Soviet leadership, Frederick is especially relevant because he activated an existing state. Under his father, Frederick Wilhelm I, Prussia was a militarized, but pacific, dictatorship. Frederick Wilhelm, a coarse and parsimonious man, carefully husbanded his resources and his manpower, but rarely used either; when his son came to power, he seized the first opportunity to strike because, as he wrote in a secret testament, "I was young, had a big army, a full treasury, and I wanted to see my name in the newspapers."

Frederick's urbane cynicism reminds one of characterizations of much of the personnel of the KGB. He was a master of propaganda and was able to convince the progressive intelligentsia of Europe that he was an enlightened ruler. As a young man, Frederick was obliged by the brutality of his father to pretend to be a dutiful son and an obedient and dull Prussian junior officer. But as soon as he occupied the throne, his true colors were displayed: he fought six wars (usually aggressively), tripled the population and land area of Prussia, and turned a secondary state into a first-line power. A less cynical and more patriotic Soviet Frederick might accomplish even more for his country.

Still, the metaphor of a Soviet Frederick is weak on two points. Frederick came to power when he was young and vigorous; Soviet leaders get to the top considerably later in life. And while Frederick was not greatly troubled by heavy casualties (in the Seven Years' War Prussia's war dead was an estimated 7 percent of the population, and the countryside was devastated), it is almost incredible to imagine an urbane cynic risking thermonuclear war to get his "name in the newspapers." Nevertheless, Andropov comes closer to an older and soberer Frederick the Great than do any of the other metaphors discussed here; but "older and soberer" are important qualifications.

WILHELM II

A more recent German metaphor for a Soviet leader is the last kaiser. Whereas Frederick was urbane, intelligent, self-confident, and cynical, Wilhelm was provincial, stupid, insecure, and high-minded—a dangerous combination. He, too, took over a state with enormous real power (power that had been wielded cautiously by Bismarck). Imperial Germany and its emperor seemed to suffer from some sort of inferiority complex that led to foolish posturing, clumsy diplomacy, and an arms program that alarmed previously friendly powers, especially England. The kaiser and his ministers, generals, and admirals succeeded in surrounding Germany with enemies. Their excessive reliance on purely military solutions to diplomatic problems helped bring about the First World War.

The kaiser metaphor applies to the Soviet Union in several obvious ways. The militarization of the Soviet Union (and its devotion to the theories of Clausewitz) is distressingly familiar. Also reminiscent is the parvenu desire of the Soviet ruling elite to be respectable and well regarded by the rest of the world. In a sense, the Soviets have already had a sort of kaiser metaphor in Nikita Khrushchev. And, it is important to note that Khrushchev's own kind of clumsy diplomacy, vulgarity, and "adventurism" led to his ouster by his colleagues.

"THE MULE"

Here we turn from historical fact to science fiction. Isaac Asimov's *Foundation* novels describe a galaxy where there is a planet of technicians who have developed a long-term plan for the survival of civilization. The plan is devised on the basis of a scientific calculation of history. But the plan is upset and the technicians are conquered by an interplanetary adventurer named the Mule. He appears from nowhere, a biological mutant with formidable personal abilities—an exception to the normal laws of history. By definition, such mutants rarely appear but they are not impossible. In a sense, we have already seen a "mule" in this century—Hitler—and another such "mutant" could conceivably come to power in the Soviet Union.

The danger of Soviet aggression and the risk of nuclear war would increase significantly if an individual akin to any of these metaphors became the leader of the Soviet Union. Yet, as I have argued, the Soviet system tends not to promote people with the kind of talent, ambition, dynamism, or craziness found in the metaphors. The Soviet Union today is ruled by a self-perpetuating oligarchy without established institutional mechanisms to force early retirement, promote turnover, or bring younger men to the fore. The highest civilian and military ranks in the Soviet Union are filled by men who have worked their way to the top through many years of self-control, accommodation, and intrigue. These men are formidable adversaries—intelligent, tough, and devoted to their system—but they are not high-risk gamblers. What they seek to avoid is nicely summed up in the Marxist-Leninist term, "adventurism."

For this reason, among others, I believe that calculated, fanatical, or stupid Soviet behavior that would deliberately or indifferently incite a thermonuclear war is, fortunately, not in the cards. However, should the present Soviet system be upset or substantially modified, I would be far less certain of this forecast, and might even begin to lose sleep over it.

PART THREE

Looking at Some Current Defense Issues

7

Mobilization in the Nuclear Era

"Mobilization" is the redirection of national resources, both human and material, away from traditional civilian pursuits to support a greatly intensified defense effort. The notion of mobilization in the nuclear era seems to be a contradiction in terms when compared with the mobilizations of World Wars I and II.

With the appearance of nuclear-armed long-range bombers, ICBMs, and ballistic missile submarines, mobilization as a concept appeared as archaic and obsolete as the military forces that in the past had been mobilized. How could national mobilizations affect the course of any future U.S.-Soviet central war if nuclear weapons could destroy most of the enemy's (and one's own) war-making power in a matter of hours or days?

Mobilization and Its Role as a Deterrent

My purpose in this chapter is to make clear that mobilization *can* play a critical role in the current U.S.-Soviet political and military competition. U.S. (and allied) mobilization capabilities can be used to "regulate" provocative Soviet behavior and "correct for" the consequences of Soviet aggression.

As I argued in chapter 6, nuclear war is unlikely to be started by a bolt-out-of-the-blue surprise attack. Instead, "protracted crisis" is the most likely prelude to war. We can imagine a generalized scenario involving a period of intense political crisis between the U.S. and the Soviet Union during which both sides fear that a nuclear war may actually occur. However, neither side is willing to risk such a conflict with the existing levels of offensive forces and strategic defenses. As a result, both nations undertake large-scale programs to develop considerably more effective strategic offensive and defensive capabilities. The competitive mobilizations taking place within the context of this kind of crisis (and its aftermath) can be called "mobilization war."

Mobilization war is warfare in the sense of a compressed and bitter acceleration of the arms competition, but it is a confrontation without the certainty that direct military action will break out. Each side would strive to obtain a superior military position in order to improve its bargaining power and to prepare for immediate or later conflict. The purpose of a "modern" mobilization is as much to influence the enemy's perceptions and calculations as it is to shift the actual prewar balance of military forces in one's own favor. A plausible outcome of this scenario is that the side that mobilizes most effectively within a relatively brief period of time (say, three months to three years) can achieve a militarily dominant position, enabling it to inhibit the diplomatic or military initiatives of its opponent.

Perhaps the closest historical parallel to mobilization war during the nuclear era arose as a consequence of the Korean War. Following the end of World War II, despite the forcible consolidation of Soviet power in Eastern Europe and the presence there of a formidable Red Army, the United States demobilized most of its armed forces and substantially reduced its military expenditures. Prior to the outbreak of the Korean War in June 1950, Congress was debating whether the defense budget should be $14, $15 or $16 billion. After the North Korean invasion of South Korea, Congress authorized $60 billion—a fourfold increase. Between 1950 and 1953, defense expenditures increased from 5 to 13 percent of GNP.

The very large defense budgets of the time were not solely dedicated to financing the Korean War. Many U.S. decision makers feared that the conflict was a Soviet-sponsored feint for an impending Soviet invasion of Western Europe. Thus, across-the-board improvements in U.S. defense were required. Total obligational authority for strategic forces, for example, increased by more than a factor of four between fiscal years 1950 ($2.5 billion) and 1952 ($11 billion), and stayed at a level of $7 to $11 billion per year for most of the 1950s.

The expansion of U.S. defense spending triggered by the Korean War permitted the development of an impressive set of military systems, including the Atlas ICBM, the Polaris ballistic missile submarine, and the Minuteman ICBM; a huge expansion of the air force (e.g., the procurement of hundreds of B-47 and B-52 bombers); an extensive air defense network; a variety of theater nuclear weapons; and modernization of elements of the navy. It is important to note that if there had been no Korean War and annual U.S. defense budgets had stabilized at around $15 billion, many of these systems would have been rejected as "technologically infeasible."

In effect, U.S. military spending in the early 1950s ensured the American strategic superiority that lasted until the late 1960s. In retrospect, it seems clear that the shadow of this stark superiority served as a major constraint on Soviet attempts in the late 1950s and early 1960s to exert military power in support of their diplomatic objectives. The Korean War mobilization was a sobering experience the Soviets undoubtedly remember.

Consequently, the Soviets probably are deterred from perpetrating a series of Koreas, Angolas, and Afghanistans, less by the fear of an immediate and direct military response than by the probability that the United States and its allies would greatly increase both their military strength and their willingness to respond effectively to such provocations in the future. The deterrent effect of such fears escalates if the U.S. and its allies are seen making explicit preparations to react to Soviet threats or provocations by an increase in military capability. These mobilization responses are usually not very dramatic, except when the mobilizing powers are faced with a dramatically unacceptable provocation or a great and credible danger of further provocation.

As important as it is to increase defense expenditures, the increase becomes much more valuable if authorizations can be translated into military strength relatively quickly. Were the Soviets concerned (and forewarned) that a deterioration in U.S.-Soviet relations would push the U.S. into a crash program to upgrade its defenses, they might be less willing to let those relations deteriorate.

If in the course or the aftermath of a serious crisis the Soviets believed that we were about to undertake a major mobilization, they would be faced with four alternatives:

1. *They could strike the United States before the buildup went very far.* U.S. Type I Deterrence should discourage this, especially since the buildup would almost certainly be accompanied by an increased alert and other measures to reduce the vulnerability of U.S. strategic forces and supporting command-and-control systems to a Soviet first strike.

2. *They could try to match the U.S. program.* If the U.S. program were large enough and effective, this would not be feasible for the Soviets, because of costs and other reasons. The sheer size of its GNP would give the United States a significant edge in a competitive mobilization with the Soviet Union. The U.S. GNP is roughly double that of the Soviet Union, and has been for some time. During his first year in office, Secretary of Defense Weinberger reportedly asked the armed services to develop contingency plans for an emergency mobilization in which half the GNP—about $1.5 trillion—would be devoted to national defense programs. One and a half trillion dollars is the sum equal to the *entire* Soviet GNP. In addition to this edge in GNP, U.S. short-run flexibility of production probably is also substantially greater than that of the Soviet Union, and this would facilitate the conversion of civilian industries to the manufacture of military products.

3. *They could increase their own budget, but still accept a position of inferiority.* Such acceptance could be a serious step, since the United States might then have a "fight the war" capability as well as "deter the war" capability. Further, the extra costs they were willing to incur might still bear very heavily on them.

4. *They could attempt to lessen the motivation for the U.S. buildup (or otherwise interfere with it) by political or other means.*

Since the costs and risks of a Soviet provocation would be increased by any of the first three alternatives, and the fourth could be risky as well as ineffective, it is likely that the Soviets would carefully calculate the cost-benefit considerations before attempting any significant provocation. If they nonetheless decided to go ahead, the U.S. could initiate a mobilization to acquire rapidly the forces to "correct for" Soviet aggression or at least to prevent the Soviets from exploiting any initial gains.

By confronting Soviet leaders with these unpleasant alternatives, a large mobilization base would provide the U.S. president with a useful option for resolving a war-prone crisis on favorable terms. Rather than ordering immediate military action against the Soviets, threatening a nuclear attack (with the risk that the threat might have to be carried out), or accommodating to Soviet demands, the president could temporize in the situation at hand—initiate a "phony war" (which might be inevitable in any case)—and rapidly, in the preplanned manner, employ the superior American technological and industrial base to build up the kind of superiority that would make accommodation by the Soviets much more probable. In a "war" of competitive mobilization, great pressures would be exerted on both sides to settle the issues at dispute short of open conflict. The United States should hold a comparative advantage in this crisis bargaining. In many cases it should be possible to terminate the mobilization war in a reasonably satisfactory manner while avoiding large-scale central nuclear war, or any nuclear violence at all. However, if mobilization failed as a deterrent, it would still have provided the U.S. with additional and more effective forces with which to wage war.

Mobilization as a deterrent interacts with the Type I Deterrence and Type III Deterrence described in chapter 5. To illustrate these interactions, let us consider the linkages among an effective U.S. mobilization base, the rapid deployment forces of the U.S. Central Command (CENTCOM), and the protection of the Persian Gulf oil fields. Any mobilization following a Soviet-induced crisis in the Persian Gulf would be triggered as much by concern for the near- and long-term Soviet threat to *all* U.S. vital interests as by the specific consequences of the crisis itself. In addition, the existence of a strong U.S. mobilization base in advance of a crisis would help to counter, if not deter, Soviet intervention in the Gulf. Depending on the nature of the attack and the tactics employed, the units initially deployed by CENTCOM could prove successful in conducting holding actions until additional reinforcements could be brought to bear on the battle. If the Soviet attack were large enough, the rapid deployment forces would almost certainly be defeated (but not necessarily with ease). Nonetheless, by compelling the Soviets to realize that they would have to destroy one or two American divisions, deployment of U.S. conventional forces would have raised the stakes of the Soviet action substantially. There would be deterrence value in any credible American resistance, especially if it were clear "up front" that destruction of the rapid deployment forces might well lead to a declaration of war or a full-scale mobilization of at least the magnitude that occurred as a result of the Korean War.

A credible mobilization deterrent requires that U.S. strategic forces be capable of deterring the Soviets during a severe crisis when they might feel great pressure to strike the United States. Knowing that an attack on the CENTCOM forces could prompt a massive U.S. mobilization, the Soviets might consider it less risky to initiate counterforce operations against our strategic missiles, bombers, and submarines than to touch off a possible mobilization. If U.S. strategic forces were clearly adequate to forestall a Soviet counterforce first strike, if we were willing to rely on this to the extent of actuating an all-out mobilization, and if this possibility were to deter a Soviet attack on the Persian Gulf (or on the rapid deployment forces), then our strategic nuclear capabilities would have played a crucial role in deterring a Soviet attack on the Persian Gulf, even without any explicit threat of their use.

The foregoing example shows how the nation's Type I Deterrence interacts with its capability for mobilization as well as its Type III Deterrence (in this case the use of the rapid deployment forces). In addition, the mobilization base substitutes for inadequate Type II Deterrence. If the Type II Deterrence were of higher quality, the Soviets could expect an immediate attack or declaration of war rather than a mobilization. But a mobilization war can substitute for appeasement (acceptance of Soviet aggression) as well as for war. Once a mobilization base has been activated, and possibly even after a formal declaration of war (but prior to an escalation of the conflict to countercity attacks), the U.S. could use its mobilization ability to gain strength for itself while perhaps diminishing the strength of its opponent, but in any case to put itself in an advantageous bargaining position. The threat of an actual U.S. first strike (or delayed Type II Deterrence) becomes quite credible after a full-scale mobilization, possibly to the point of putting so much pressure on the other side that it either compromises or preempts. (As mentioned above, a compromise seems more likely since mobilization means that Type I Deterrence has also been increased.)

The idea that mobilization is a way of punishing provocations (in addition to preparing a country for waging war) also applies to the deterrent value of mobilization by U.S. allies. Soviet fear of creating an overwhelming worldwide coalition against them also helps moderate their aggressive tendencies (although the threat of Soviet aggression may remain at a dangerous level). For example, unless the Soviet Union really wanted to go for broke and take over Europe, it would probably not be willing to violate the nuclear threshold in any limited crisis. The Soviets would know that doing so, even if it yielded limited gains, would almost guarantee an enormous arms buildup against them by Western Europe, the United States, and much of the rest of the world. (A buildup in the "rest of the world," in this case, would include the likely rearmament of Japan and perhaps sizable U.S. military assistance to China.)

The Soviets would have to heed the classic warning to opponents of the established order: When you strike a king, you must strike to kill. Unless their (brutal and unequivocal) use of nuclear weapons were to change dramatically the world balance of economic and technological power in their favor, they would almost certainly find most of the world arrayed against them. As a result, certainly in the long run, the use of nuclear weapons would have been a mistake. Think of the impact of a strongly armed and hostile China and Japan, a massive mobilization program by the United States and/or Western Europe, and perhaps greatly increased U.S. and/or West European arms transfers to other nations. The Politburo would have to consider this latent anti-Soviet front a powerful deterrent, especially if there were a political and ideological "mobilization base" to match the technological and economic one. At least to some extent, these possibilities affect the credibility, likelihood, and character of any potential Soviet aggression—particularly one involving the use of nuclear weapons or an attack on an especially sensitive area such as the Persian Gulf.

The degree of provocation and the degree of mobilization are variable but interrelated. Actuation of a worldwide anti-Soviet mobilization base would require fairly unambiguous events and prospects—most likely including a succession of serious disruptions of the status quo through the use or brandishing of armed force. Even then the emergence of a broad and resolute defensive coalition could not be guaranteed. While Hitler's actions in the late 1930s (e.g., the occupation of the Rhineland, the *Anschluss* with Austria, the Munich crisis) were about as unambiguous as one can reasonably expect, a high level of allied mobilization did not occur until after Munich. Similarly, the Soviet role in Afghanistan and Poland has elicited only mild anti-Soviet counteractions around the world. (Yet, were the Soviets to intervene in Poland more directly and were the intervention to grow long and bloody, then the horror of the events themselves would probably mobilize world opinion against the invaders.)*

The point here is that to the extent hostile activities can be viewed

---

* However, the threat of such mobilization would probably not be sufficient to deter the Soviets if definitive intervention were the only way of preventing Poland from becoming a neutral or hostile power. All the main Soviet communications and transportation links with East Germany pass through Poland; if those links were severed or even became precarious, the security and effectiveness of Soviet forces in East Germany, as well as the political and military balance in Europe, would be drastically altered. It would be almost inconceivable for the Soviets to allow the Poles to do this. In fact, we could almost argue that very few West Europeans, including very few West Germans, would be happy to see Soviet control of East Germany made so uncertain.

as ambiguous or inevitable rather than as directly threatening, an aggressor's actions will not bring about a particularly dramatic worldwide response. For example, many minimize the danger to international peace and security posed by the invasion of Afghanistan by arguing that the Soviets felt compelled to intervene to stabilize a chaotic situation on their periphery and to discipline a recalcitrant satellite. (The real takeover, it should be noted, had actually occurred a year and a half earlier.) According to this argument, the Soviets were drawn into a military adventure that turned out to be a serious mistake; the intervention revealed no *new* Soviet malevolence; and the occupation of Afghanistan has actually weakened Soviet military capabilities.

In short, worldwide mobilization is activated in proportion to the level of provocation or threat. It is not a weakness of the worldwide mobilization *concept* that stronger actions have not been taken to counter Soviet involvement in the tragedies in Poland and Afghanistan. Rather, it is an indication that most of the rest of the world has not felt unduly provoked or directly menaced by the Soviet actions. If a provocation can essentially be construed as a Soviet "domestic matter" or one in which the Soviets have clearly limited themselves (as in Poland), a global outcry is unlikely to ensue. This does not mean that the aggression should go unpunished or unnoticed; the Soviet Union should be subjected to penalties, in part as a matter of justice and good policy, in part to raise, in a demonstrable way, the level of the West's sensitivity to future provocative acts. For this reason, it is important that mobilization bases be capable of effective response at relatively low levels of aggression, as well as those that precipitate the emergence of massive countervailing power.

Up to this point I have used the word "provocation" to refer to blackmail backed up by military force, territorial aggrandizement, and certain kinds of limited war. However, Soviet violation of an arms-control agreement would also constitute a "provocation." Mobilization capabilities can reinforce arms control, particularly in making so-called "breakout" efforts unprofitable. (Breakout involves the relatively sudden, explicit, and unilateral abrogation of an arms-control agreement through the fielding of new or additional weapons that had been constrained by the broken treaty. With the illegal deployment, the violator hopes to gain significant military advantages or political leverage.) If a nation can react quickly to violations of an arms-control agreement, or even just a fundamental—yet "legal"—change in the strategic balance, that nation will be much more secure in accepting agreements with some potential problems (like breakout) than if it did not have a mobilization base with which to enforce the arms limitations. Thus, maintenance of a mobilization base can help the United States in pursuing its arms-control objectives in peacetime, deterring provocations that disrupt the international order, resolving crises on terms consistent with our security interests, and waging war successfully if deterrence breaks down.

U.S. Mobilization Capabilities

The concept of mobilization war in the nuclear era implies relatively short reaction times for the deployment of major offensive as well as active and passive defensive systems which may be extremely complex and costly by peacetime standards. During a large-scale mobilization, it is entirely plausible that the U.S. strategic budget alone could constitute an expenditure of several hundreds of billions of dollars per year. Expenditures of this magnitude would make possible a very wide range of military and nonmilitary* defense systems that could not be seriously considered within recent strategic budgets, which have amounted to between $10 and $30 billion annually. For example, a multilayered ballistic missile defense (BMD) system, employing lasers, particle beams, and other advanced techniques for boost-phase, midcourse, and terminal interception of enemy warheads, could, in principle, be procured under conditions of mobilization war.

The crucial determinants for acquiring such a capability lie in the prior research and development (R&D) program and in a proper institutional orientation within the government toward the nature, role, and implementation of mobilization. The requirements of a mobilization base to support mobilization war are sufficiently different from the objectives of existing R&D programs needed to meet current and near-term defense requirements that funding for a mobilization base should be partitioned from other R&D expenditures.

The primary function of a mobilization base is to compress the lead times needed to acquire highly effective strategic forces, nonmilitary defenses, and conventional forces within the relatively short periods of time associated with mobilization. Under some circumstances, it may

---

* "Nonmilitary defense" refers to measures to provide the population and economic infrastructure with direct protection against nuclear attack. This would include civil defense, ballistic missile defense, and air defense programs.

be sufficient to have "paper plans," say, for the conversion of designated industrial plants from civilian to military production. In other cases, where the military requirements are more stringent and less easily adaptable to short-term changes, some limited development or prototyping of weapon systems or defense measures may be necessary. In still other cases, particularly where the function is highly complex and likely to involve large numbers of both civilian and military personnel, such as a BMD or civil defense system, it may be necessary to carry out a limited deployment or field testing in peacetime, and to develop the professional cadres who could support a vast expansion in the system during a mobilization.

It is sometimes argued by defense analysts and economists that the United States no longer has the ability to duplicate its mobilization efforts of World War II and the Korean conflict. They argue that the economic system has become too rigid, that a rapid change from civilian to additional military production is not possible, and that our increasingly service-oriented society provides neither the appropriate capital resources nor the skilled manpower needed to make large-scale mobilization a credible U.S. defense option. They also assert that weapon systems today are vastly more complex and difficult to produce, and that new systems could not be developed within the time frame (three months to three years) required for mobilization.

These critics greatly underestimate both the flexibility of U.S. productive capacity and the technical resources and ingenuity of U.S. industry. There is currently substantial slack in the U.S. economy that could be picked up during a mobilization, and economic resources could always be converted (either voluntarily or through emergency legislation) to military production. For example, any new building construction that has not been started (or is less than 5 percent completed) could be halted. The production of civilian automobiles could be severely curtailed or shut down. Most construction and manufacturing could be limited, thereby forcing many industries to participate in the mobilization. On the other hand, the government could reward voluntary participants for early and intense involvement in the mobilization effort, thereby creating a situation where industry would be anxious (if not desperate) to rise to the occasion. Clearly, all efforts would depend upon the degree of the emergency and the extent of action by the government.

But what about the feasibility of actually converting industry so that it could rapidly manufacture appropriate military products? The usual mistake in estimating U.S. mobilization potential is to ask what can be done using customary techniques under customary pressures. In normal periods if a manufacturer has 10 to 20 percent higher costs than his competitor, he is likely to be in serious trouble; this is also true if he is in the business of producing military hardware and his performance is 10 to 20 percent less than his competitor. In a mobilization, the government can afford to subsidize inefficient production, at least initially, and may also be willing to compromise on the performance of weapon systems. The military could modify many specifications and thereby increase the number of potential contractors.

Assessments of U.S. mobilization capabilities should, therefore, not assume business-as-usual conditions. There is a wide range of strategies that could be adopted in a crisis to facilitate mobilization and ensure rapid and effective production of military materiel. Some of these strategies are identified below, along with examples of their applications.*

Change manufacturing techniques. Manufacturing techniques used to produce military equipment can be altered to take advantage of available resources or to overcome production bottlenecks. There are many opportunities to "work around" bottlenecks by downgrading the specifications for military equipment. Lowered specifications could yield a large increase in production rates, with only a modest decrease in combat performance. Work-around procedures were common practice in previous wartime industrial mobilizations by the United States. Many current analyses of U.S. mobilization potential, with a focus on preserving the highest levels of system performance at minimum operating and maintenance costs, tend to overlook these historical precedents.

A potential application of the work-around concept can be found in the manufacture of tank turrets. In one study, the Hudson Institute examined the ability of the U.S. to increase rapidly its production of M-60 main battle tanks. The turret for the M-60 was cast in one piece. While the structural integrity of the turret was excellent, casting it in one piece restricted its manufacture to only a few foundries. This created a latent production bottleneck. We concluded that one way to circumvent the bottleneck would be to make the turret of two or three smaller pieces of cast steel that had been welded together. With this change in manufacturing technique, the number of foundries capable of

---

* Many of my conclusions concerning the mobilization potential of the U.S. economy are based on several Hudson Institute studies, notably those of my colleague Frank Armbruster.

producing the turret could be expanded significantly, although the cost per turret would be higher and there would be a slight (tolerable) degradation in the strength of the turret.

Find additional sources of supplies. The frequently underestimated flexibility of the U.S. economy becomes very important in a situation where components cannot be obtained readily from standard or customary sources of supply. Hudson Institute interviews with aircraft company officials indicated that a major bottleneck in the increased production of tactical aircraft during a mobilization would be a lack of the large forges needed to manufacture main landing gear components and wing center sections. The aircraft company officials believed that only one or two U.S. firms could produce the necessary forgings. However, later interviews with representatives of the forging industry revealed that a number of other firms could produce the landing gear and wing parts.

Change materials or other component specifications. A shortage of materials can often be rectified by substituting others, sometimes without making any changes in design. In the late 1950s, the Rand Corporation conducted several studies of postattack recuperation. In the course of this research, it was discovered that the inability to reconstruct piers would be an impediment to economic recovery. The unavailability of the special huge timbers used in the construction of the piers was the constraint here. It appeared that one remedy for the problem would be to substitute rocks for the scarce timber (at about a 10 to 20 percent increase in cost).

Similarly, most metals could be interchanged, albeit with necessary modifications to bulk and cost.

Change design. A change in the design of equipment is another means of quickly expanding military production in a mobilization. A design change or substitution could be made for gas turbines in tanks or warships. Since gas turbines have a very good power-to-weight ratio, they are absolutely essential in aircraft. But since the power-to-weight ratio is much less important in tanks or warships, these are both cases where gas turbines could be replaced by diesel engines (which the United States can easily make in large quantities).

Soviet experience provides another example of this concept. The MIG-25 is a high-performance interceptor. However, the plane gets into trouble at about Mach 2.8 (a speed within its maximum limit). Had the MIG-25 been an American design, it probably would have been equipped with a complicated and costly warning system to alert the pilot to the danger of flying at Mach 2.8, or even to prevent the plane from exceeding that limit. The Soviets simply put a red warning line on the Machmeter, not bothering with any elaborate fail-safe system. This example illustrates the possibility of accepting a loss in performance (i.e., a combat speed less than the maximum speed) and a simple fix (i.e., the red line). Yet, the MIG-25 has always been considered a high-performance aircraft, even though not as advanced technologically as it might have been—and as many American observers, impressed by its performance, thought it was until an example (provided by a defector) was inspected.

Procure from "nondedicated" sources. Many argue that the U.S. ability to mobilize is limited by the size and efficiency of our peacetime (or "dedicated") defense industry. But the U.S. mobilization base encompasses a much broader portion of the economy than just the current defense sector. The military buildups for World War I and II involved the two largest industrial mobilization efforts by the United States. In both cases there was extensive use of industrial facilities not previously dedicated to military production. Nondedicated industrial potential was rapidly exploited, with little prior planning.

For example, General Motors built more Grumman-designed Wildcat fighter aircraft during World War II than Grumman did. Goodyear Tire and Rubber Company turned out most of the Marine Corps' FG-1 Corsairs, not Chance-Vaught (the traditional airframe manufacturer). Automotive companies and the firms building railroad locomotives and cars converted to the large-scale production of tanks for the war effort. Despite differences between the U.S. economy of the World War II period and that of the present, I would still expect to see in any future mobilization comparable examples of ingenuity and flexibility in the conversion of the enormous nondedicated industrial sector to military production.

In a mobilization effort intended to provide the population with protection against nuclear attack, the U.S. construction industry could be redeployed to build blast and fallout shelters and to "harden" various economic assets critical to postattack recovery. Almost any firm experienced in construction could, given the blueprints, start building shelters almost immediately. These shelters might not be built at a competitive cost, but in the intense crisis creating the context for the mobilization, competitive costs would not matter much. One could imagine the bulk of such a nationwide shelter program being completed in a year or two, if the preliminary plans and bureaucratic red tape had been taken care of before the decision to mobilize.

Nondedicated sources of supply might also be nonnational sources. While domestic sources are preferred in most cases because dependence on foreign sources risks vulnerability to external disruptions, there has developed over the past thirty years an enormous worldwide industrial base upon which we could draw. The U.S. now represents only about a third of world industrial output (excluding the output of the Soviet Union, Eastern Europe, and China). Brazil, Mexico, Japan, South Korea, and Taiwan (among others) might provide important contributions to a U.S. mobilization. The critical aspect of foreign sources would not be the absolute level of the inflow of goods, but rather their ability to break production bottlenecks. The development of an effective offshore mobilization base might be aided by the judicious encouragement of joint production contracts with foreign governments and firms.

It should be pointed out that the existence of a large and flexible nondedicated industrial mobilization base does not mean that peacetime defense industries are unimportant. Dedicated military production facilities are central to the full exploitation of the latent mobilization capabilities of U.S. industry as a whole. The dedicated production base has the managerial, scientific, and industrial cadre that would be crucial in shifting civilian plants to the manufacture of war materiel.

Change military tactics (including changes in military forces). Changes in military tactics may be called for if certain additional military equipment cannot be procured in the required time. In the course of a mobilization, for example, NATO might revert to laying mine fields and establishing other antitank barriers to defend against a Warsaw Pact blitzkrieg on the Central Front. The U.S. and its NATO allies currently are developing complex weapon systems, based on sophisticated technologies, for attacking the armored forces of the Warsaw Pact. If sufficient numbers of these systems could not be deployed, even under mobilization conditions, large quantities of land mines could instead be produced. Mines are relatively simple to turn out, and mine fields can be laid by a variety of relatively quick and easy methods. Mine fields would slow down the enemy's armored thrusts (disrupting his timetable), while NATO mechanized units would be employed as mobile reserves to counter breaches of the defensive line.

Change military strategy. This is similar to changing tactics in that it involves altering plans to coincide with new limitations or new advantages that result from mobilization. But it could also create a more fundamental or far-reaching change in the U.S. approach to a conflict. A full-scale strategic mobilization might enable the United States to shift its nuclear strategy to a greater concentration on active and passive defenses against a Soviet strike. We have already mentioned that a mobilization might result in the deployment of a wide-coverage BMD system and a civil defense shelter program.

Exploit enemy and/or other uncertainty and ignorance. To the extent that one is trying to impress the enemy and one's allies, it may be possible to utilize programs that are more or less "fake" but look real or can be rapidly improved. (Observers tend to credit weapon systems with their design capabilities even if it takes time to attain them.) This is particularly useful if security is maintained or remote observation will not permit the enemy to be certain if the programs are "fraudulent" or not. In many instances these sham or early deployments can eventually be given real utility, but their initial purpose is psychological.

To a remarkable degree, the Luftwaffe of the mid-1930s was the military equivalent of a Potemkin village. Prior to World War II, Hitler inflated its actual capabilities for strategic bombing in order to frighten his neighbors into making major political concessions. Khrushchev likewise rattled his embryonic rocket force in the 1950s to intimidate the West during the Suez and Berlin crises.

Use educated guesses. When mobilization must be achieved as quickly as possible, and when victory or defeat depends on the rapid production of needed weapon systems, it may not always be possible to conduct the rigorous research, development, testing, and evaluation that are standard in the development and deployment of new systems. In such cases, expert intuition, incomplete studies, "learn-as-you-go" projects, and parallel experimental programs will have to compensate for a lack of information or experience.

Establish a premobilization R&D program. A comprehensive mobilization strategy implies the creation of a resource and planning *base* in peacetime, from which a bigger and more specific mobilization program can emerge if a crisis occurs. Prototypes—design models that can later be put into production—are an important component of a mobilization base. They are physical blueprints for a weapon system that can be built in large numbers if necessary. Research and development is crucial not only for the design of new weapons, but also as the basis for making responsible choices between alternative projects and systems. By spending a few billion dollars on a premobilization R&D program, the government would be in a much better position to spend effectively the additional tens or hundreds of billions of dollars for defense programs that would be authorized during a mobilization.

Study of the clandestine rearmament of Germany after World War I provides many examples of the contributions R&D investments can make to the rapid expansion of a nation's military strength. Research and development was conducted along lines that supported the conceptions of the German General Staff concerning the character of future warfare, particularly the employment of relatively small (by World War I standards), but highly mobile, armies with extensive use of air power. The actual construction of prototypes and their use in maneuvers or in foreign theaters of operation provided a technical basis for drastically cutting lead times for the development of high-performance mechanized equipment and aircraft. In addition, the clandestine prototyping of other weapons, such as field artillery, antiaircraft artillery, and small arms, maintained the viability of an R&D community for the larger exploitation of available skills and manpower in the late 1930s, when Germany's rearmament became overt.

Use preplanned "stop" and "trigger" orders and policies. As noted, the civilian economy can be mobilized effectively by the use of carrots and sticks; that is, by decreasing the market's ability to get materials for industries and activities that are no longer thought of as high priority and increasing the opportunities for industries and activities that are. It should be realized, however, that because mobilization is not quite as drastic as war, market forces should, where possible, play a major, even dominating, role (though there may be many controls and allocations as compared to business as usual). Even so, this will clearly require sensible financing, control of the money supply, cutbacks in some social welfare programs (made less onerous by nearly full employment), and so on, and presumably a substantial increase in taxes or government borrowing.

Exploit many other flexibilities of the U.S. and other economies. The nondedicated plant production and work-around procedures for mobilization could be supplemented by other innovative industrial techniques. These include CAD/CAM and the vertical breakdown of production processes to play to the comparative advantages of domestic and foreign manufacturers. CAD/CAM stands for "computer-assisted design/computer-assisted manufacturing" and represents a genuine technical breakthrough. Indeed, it is perhaps the most spectacular example of technology that can make manufacturing incredibly flexible, enabling the production of new kinds of equipment and the implementation of all kinds of changes rapidly and efficiently—i.e., making future mobilizations easier than their historical counterparts. Reduced manpower (at all levels of skill) can be offset by greater use of the reprogrammable robots, automated equipment, numerically controlled machines, and so on, just now seriously coming "on stream" in manufacturing.

Some Caveats and Final Notes

Mobilization is not a panacea; it should be clear by now that it will not serve as a replacement for strong existing military forces, nor will it be sufficient by itself either to deter a wide range of Soviet aggression or resolve intense crises. Mobilization is not a low-cost solution to the nation's defense problems (as in "Why build—and pay for—large military forces now, when for much less money you can prepare for building them when they are needed?"). It would be a serious misapplication of the mobilization concept if it were to affect adversely the present commitment to strengthening and modernizing U.S. forces-in-being. Strong current forces are vital to any defense planning, including plans which emphasize mobilization. As noted earlier, mobilization capabilities reinforce the deterrence value of nuclear and conventional forces-in-being, while those forces, in turn, allow time for an activated mobilization base to influence the course and outcome of a conflict. U.S. strategic nuclear forces must have a war-fighting capability as the best deterrent to a Soviet preemptive strike designed to achieve victory before a mobilization can occur. (A war-fighting posture is also the best protection in the event a war has to be fought immediately.)

The U.S. has historically demonstrated a tendency to neglect its defenses during times of peace and (relative) international tranquillity. Increases in defense spending must often be justified in terms of immediate threats to U.S. security. Thus, for example, the Reagan administration has tended to associate its rearmament efforts with what is almost a renewal of the cold war—or at least an antidétente policy. This linkage has increased international tension and created a backlash (partly motivated by increased fears of war) against the spending program. U.S. defense planners must recognize that the American public's (to some degree justifiable) aversion to military preparedness is a basic characteristic of U.S. "strategic culture," and plan accordingly. One of the advantages of a mobilization program is that it can provide a much more acceptable bridge between current force levels and those that might be necessary to meet a serious crisis or conflict with the Soviet Union. This would allow U.S. defense planners to be more sanguine about gaps in current U.S. capabilities and enable them to work to fill these gaps without creating an often unproductive, or even counterproductive, atmosphere of tension and hostility.

Opponents of mobilization fear that a strong mobilization base will make U.S. political and military leaders overconfident and encourage either rigid or reckless behavior in crisis situations. They argue that the knowledge that U.S. mobilization capacity is likely to give the U.S. a significant military advantage might prevent its leaders from compromising in crisis negotiations, thus possibly forcing the Soviets into a corner from which nuclear war appears to be the least disadvantageous alternative. A corollary to this argument holds that the U.S. might deliberately seek out crises in order to justify increases in its military forces.

A more realistic caveat for mobilization planning involves the tendency to underestimate the Soviet will and ability to counter any U.S. mobilization effort with a similar effort of their own, regardless of the economic and social cost. While the details of Soviet mobilization potential are largely unknown, we do know that such an endeavor would not be easy for them—the Soviets already devote a much higher portion of their GNP to defense than does the U.S., and their economy is less flexible than its American counterpart. However, under the Soviet authoritarian political system, its leaders can arbitrarily compel sacrifices and reorder priorities to an extent that would not be possible here.

U.S. alliance partners might choose to disassociate themselves from a U.S. mobilization effort, depending on the nature of the crisis and the *perceived* risk to their own national security (or even survival). In addition, since the international economic ramifications of a major crisis plus mobilization are not well understood, one cannot rule out large or long-term dislocations in the international economic system. If our allies are unwilling to risk such dislocations, it could have profound effects on the ability of the U.S. to mobilize.

On the domestic front, an extended period of tension during which armed forces (including nuclear forces) would be on constant alert would probably interfere with a mobilization effort. The tension could result in major pressures on the U.S. government to dispense with a buildup to resolve the situation, even at the expense of capitulation. If the U.S. resorted to population evacuations during a protracted crisis, mobilization-related production in urban areas would be interrupted. In the best-case scenario, a crisis would create a national spirit of unity and self-sacrifice, with people rallying to overcome difficulties and solve problems. But an extended crisis could place enormous strains on the U.S. economic and political system, making the attainment of mobilization goals more problematic. Cooperation might not be total, especially if the crisis was seen to be the result of U.S. diplomatic rigidity, or if the threat of nuclear war suddenly became very real. The nature and handling of the crisis, as well as its duration, would be very important factors in determining what kind of response is provoked in the American people.

These are all important issues, which admittedly require further study. One of the most serious deficiencies in U.S. defense preparations has been the inadequate consideration given to the role of strategic mobilization. Nevertheless, in the nuclear age, mobilization is a source of American power and influence that if properly planned for and employed could have great positive effects in deterring provocation, waging crises, and fighting wars.

8

Civil Defense

Until the development and perfection of thermonuclear bombs in the mid-1950s, aerial bombardment of urban-industrial concentrations in wartime could be justified as an attack upon strategically vital assets: the enemy's war-supporting industries and manpower. Civilians represented a second kind of defense that provided men, materiel, and morale for the fighting forces. The same doctrine of strategic bombing that made cities high-priority targets during World War II also justified civil defense as a military necessity. Civil defense measures contributed directly to saving the lives of essential skilled workers and potential soldiers; they reduced the impairment of war production resulting from the destruction of workers' homes; and they supported the morale of those at the front by protecting those at home.

Today, civil defense is correctly perceived as contributing next to nothing to the achievement of traditional wartime military objectives. Moreover, its contribution to our primary objective of avoiding war is of secondary importance. Even the most extensive civil defense measures could not reliably hold casualties and property damage to levels that would allow an amount of postattack war production large enough to affect the outcome of an ongoing military conflict. It is prohibitively difficult to provide a great degree of civil defense protection to populations subject to direct and massive attack, particularly a surprise attack (where the option of evacuating cities would be foreclosed).

Thus, the present case for civil defense does not rest on the direct contribution it can make to a war effort. Nor does it rest primarily on strengthening our ability to deter a Soviet attack. After all, our military forces exist to protect our people, not vice versa. The main question, then, is whether, in future circumstances that we must assume are not unlikely, feasible civil defense measures can provide a material degree of protection to lives and property, and facilitate recuperation after a war is over. The answer to this question is undoubtedly yes.

To be sure, military forces can contribute indirectly to saving lives, protecting industries, and improving the prospects for postwar recovery. Counterforce weapons, survivable command-and-control systems, and plans for limited nuclear war can limit damage by lessening the scale of an enemy's retaliatory strikes and providing him with incentives to restrict the scope of his attacks to military, rather than civilian targets. But our strategic offensive forces need to be complemented by an intelligent civilian protection program and with preparations designed to cope better with the medical, economic, social, and political problems that would arise in a postwar world. Indeed, to the extent we evaluate our overall military posture by its ability to restrict damage to our people and property in the event of a nuclear war, it becomes clear that civil defense needs an increased emphasis.

I do not urge additional civil defense measures as a first-priority national-security objective. Obviously, the first priority must continue to be the deterrence of nuclear war. But even lesser priorities can be important. I recognize that any particular civil defense measure ultimately may prove to be useless, either because it was not needed or because it was overwhelmed by an attack. But there is always a chance that our efforts to avoid war may not be successful. In that event, civil defense measures are likely to be at least partially effective against most of the attacks we can envision and very effective against some possible attacks. We have a compelling obligation, at once moral and political, to examine and implement the kinds of steps that might greatly reduce war-related deaths, destruction, and human suffering—whether or not we can rely on such steps completely.

The Effectiveness of Civil Defense

Opponents of civil defense contend that there can be no effective defense against nuclear attack. Therefore, a program of "crisis relocation'' (i.e., preattack evacuation of the population to rural areas unlikely to be targeted by the Soviets) and other preparations for survival would do nothing but waste money and energy that might be better spent on social welfare programs, or, from another point of view, on improving our nuclear deterrent or our capability for waging conventional warfare. If pressed, most opponents do not question the inescapable technical facts that very moderate measures would save many millions of lives if a Soviet attack were limited to strikes against our military installations, or if it were not directed at civilians until late in a war, or if the attack were somehow less damaging than a worst-case analysis would indicate. Antinuclear activists (like the Physicians for Social Responsibility) dwell, instead, on the types of attacks for which civil defense plans would be relatively useless; for example, a surprise city-bursting attack carried out in the most malevolent manner. They typically discuss the consequences of such an attack by describing in graphic detail what a multimegaton weapon would do to the unfortunate residents of a major U.S. city, say, New York.

Of the many possible sizes and shapes of a nuclear attack, none can be ruled out, but some would be judged more likely than others. The worst possible kind of attack—a massive surprise attack directed against population centers—presents a virtually impossible problem of protecting the people in the target areas. But our weakness in the worst case does not settle the issue. For one thing, as we discussed at length in chapter 6, such an attack is one of the least likely possibilities. Moreover, civil defense programs designed to meet somewhat less ferocious and more likely wars (e.g., limited counterforce attacks that aim only or mainly at military forces and facilities) could save many lives even in the worst or less probable cases.

Before we can estimate how many people could survive in a nuclear attack, we must at least know: (1) what the targeting of the attack is; (2) whether the warheads detonated are ground bursts or air bursts; (3) how many total megatons are exploded; (4) what, if any, civil defense precautions have been taken; and (5) how much warning time preceded the attack. Of these variables, we can have reliable control only of preattack civil defense measures. But that does not mean that the others will take on the worst possible values or that we cannot influence them at all.

Probably the most important variable in estimating survival with or without civil defense is the targeting. In a "counterforce-plus-avoidance" attack (an attack carefully designed to inflict military damage while minimizing to the maximum extent possible civilian casualties and property destruction), the death toll could be relatively light even with no civil defense; whereas in a countervalue attack (one directed at population centers) the toll would be heavy, even with a most impressive civil defense program.

There is good reason to believe that almost any Soviet first strike would be aimed at U.S. strategic forces rather than American cities, and in fact cities might be spared deliberately (even to the extent of decreasing the damage inflicted on military targets). Counterforce targeting would be most sensible for the Soviets, who would then be in a better position to use the cities and the attack survivors as hostages to constrain any U.S. retaliatory strike and negotiate a favorable cease-fire.

The ratio of air-to-ground bursts used would probably make the second greatest difference. Only ground bursts produce the local fallout that can threaten people even hundreds of miles from the explosion unless they have adequate fallout shelters. Since the make-shift shelters called for by the present civil defense program would provide fallout protection, it is clear they would have relatively little value in the event that air-burst weapons were used against population concentrations.

Variations in the total megatonnage, somewhat surprisingly, do not seem to affect the toll nearly so much as variations in the targeting or the type of weapon bursts. A large increase in megatonnage against missile sites or airfields distant from urban areas might or might not cause tens of millions of casualties, but even relatively little megatonnage aimed at population centers would cause the same, if not even more, catastrophic damage.

Many assume that because of the speed of ICBMs, any attack would come as a complete surprise. Such an attack is clearly possible but by no means likely. A central argument in our discussion of mobilization was that the international tensions giving rise to an attack are likely to provide some strategic warning. The precise moment of the attack might not be known, but the nation might be in a heightened state of readiness with its cities at least partially emptied (either through spontaneous action or a preplanned and organized evacuation ordered by the government).

Variations in the above variables would greatly affect the number of fatalities the United States would suffer in a nuclear attack. Counterforce attacks might cause from 1 to 20 million immediate deaths. Attacks against economic as well as military targets might yield anywhere from 20 to 160 million fatalities.* For both counterforce and more comprehensive attacks, civil defense measures (especially crisis relocation) could save tens of millions of American lives.† While these fatality estimates represent catastrophic human losses, the differences between counterforce and less constrained attacks, and a protected and unprotected population, are significant, to put it mildly. Our efforts should be directed toward holding casualties to the lowest level possible, across the full range of potential attacks.

U.S. intelligence asessments (including unclassified reports by the Central Intelligence Agency and the Defense Intelligence Agency) indicate that the Soviets are making substantial preparations to protect their political leadership, industrial infrastructure, and citizenry from the effects of nuclear war. Their civil defense measures include in-place blast and fallout shelters, planning for the evacuation of cities during intense international crises, and procedures for the rapid construction of improvised shelters. Measures also apparently exist for the preservation of certain industrial assets and postattack reconstitution of vital sectors of the economy. Although the effectiveness of such precautions would depend critically on the military and political specifics of a nuclear war, most studies conclude that under most circumstances, evacuation and appropriate sheltering of the Soviet population could reduce casualties from 10–100 million to 2–20 million. As discussed later in this chapter, Soviet civil defense, by decreasing the likely or potential damage to people and industry, could significantly bolster the credibility of pre- and intrawar Soviet nuclear threats.

The Potential for Postattack Recovery

One frequent argument against civil defense measures is that while they unquestionably would be effective to some degree, they are still not worth taking because the long-term consequences of a nuclear war would make life impossible for the survivors. Now it is true that

* See Office of Technology Assessment, *The Effects of Nuclear War*, (Washington, D.C.: GPO, 1979), and Bruce Bennett, *Fatality Uncertainties in Limited Nuclear War* (Santa Monica, Calif.: Rand Corp., 1977).

† See "Testimony of Samuel P. Huntington," in *Civil Defense*, Hearings before the Senate Committee on Banking, Housing, and Urban Affairs (Washington, D.C.: GPO, 1979).

for many years the postwar environment would be more hostile to human life than the preattack one. However, objective studies indicate that this environment would not be so hostile as to preclude, at least in the long run, decent and useful lives for the survivors and their descendants.*

One must recognize that for most people deep grief is alleviated by time, that people go on living after tragedies, that life does go on. In fact, it is possible that most survivors would not go through as horrible a set of personal experiences as many Russians, Germans, Poles, Yugoslavs, Japanese, and others did in World War II. These people have been left with deep emotional scars, yet few of them now feel that they would be better off dead; most are leading "normal" lives—lives that by and large are indistinguishable from those of their neighbors who have suffered only "ordinary" peacetime tragedies.

In addition to psychological disruptions, a much reduced standard of living would be one of the consequences of a large nuclear war. We must remember, however, that our standard today is far higher than that required for the mere preservation of life. The present high standard of public health in America could be lowered without catastrophic consequences. For example, though much of the food grown postattack would be radioactively contaminated to an extent that would exceed peacetime standards, most of this food could be consumed without causing serious illness. This threat could also be considerably reduced by appropriate countermeasures.

Under anticipated conditions, individuals would run somewhat greater risks of various diseases and of genetic damage, but these risks would not be statistically a great deal larger than those accepted today. Consider one threat that horrifies people when confronted with the possible consequences of a thermonuclear war: the percentage of children who would be born with serious defects because of their parents' exposure to radioactivity. Following a war big enough to have irradiated child-bearing survivors with an average dose of 200 to 250 roentgens (a relatively high but not unreasonable estimate of radiation exposure among the survivors, assuming modest precautions against fallout, in a relatively destructive but not unlikely war), the present level of approx-

* See the sources cited in the footnotes on pp. 181 and 182, and a series of studies on disasters and their aftereffects by the Disaster Research Group of the National Academy of Sciences–National Research Council. The assertion that people could recover from the personal and social trauma of a nuclear war neither advocates such suffering nor views it callously, but merely tries to make an accurate assessment.

imately 4 percent born with serious defects might increase to an estimated 5 percent.*

It can, of course, be said that this is an intolerably large increase; in
any event it would most certainly be deplorable. Some will assert that
even one more deformed infant is too many. One can hardly disagree.
Yet the fact remains that life would go on. People have lived under far
worse conditions than we in America are accustomed to today, not only
throughout most of human history, but even in vast areas of the contemporary world. To argue that an effort to save people's lives is useless
because postattack life would be harsher is tantamount to saying that it
is preferable for people to forfeit their lives than to endure a lower
standard of living or of health. It is hard to think of any other equally
preposterous proposition that serious men are willing to back. Actually,
most of those who hold this view would probably find, upon introspec-
tion, that they were really comparing peacetime with the aftermath of a
nuclear war and finding they prefer the first environment. So do I. If
only the choice were that simple!

Still another argument is that a nuclear war would mean the end of
democracy as a political system. The problems of political trauma and
recovery are difficult to deal with. Americans live in a very stable country. Ours is one of the few countries in the world in which the government does not regard internal revolution and subversion as an urgent
concern. However, such problems might well exist in the postwar
world. Even if we won the war, it is conceivable that we might lose our
democracy. But again, if adequate preparations can be made, our democratic institutions and traditions can probably survive most kinds of
thermonuclear war. For some relatively small wars this is almost certainly true; for others, this is a judgment based on the belief that while
the lives and thoughts of all the survivors would be affected by the war,
their character structure and value system would probably not be so
changed as to cause them to prefer a totalitarian to a democratic system
—though there might well be a temporary suspension of many of our
customary and normally vital democratic procedures. It might be added
that the survivors of a nuclear war would be much more apt to rebel
against a democratic government that had failed to foresee the major

* Here it is worthwhile to point out that surveys of Hiroshima and Nagasaki survivors
and their children have failed to confirm the existence of damaging genetic effects. See
Committee for the Compilation of Materials on Damage Caused by the Atomic Bombs in
Hiroshima and Nagasaki, *Hiroshima and Nagasaki: The Physical, Medical, and Social
Effects of the Atomic Bombings* (New York: Basic Books, 1981), pp. 319–20.

problems of the postwar world and failed to provide those prudent efforts required for protection and recuperation. Governments unable to make satisfactory adjustments to the hard realities of the nuclear age, they would reasonably argue, are simply unfit to survive. There is some merit to such an argument.

Those who maintain recovery from nuclear war would be impossible or unlikely often portray the United States or the Soviet Union as one huge interdependent organism with many vital and vulnerable portions. Each part is assumed to be crucial to the proper functioning of the whole. This picture is largely misleading. It is true that each country does have vital sectors that are necessary for smooth operation in the short term, but in the long run, each country would be remarkably resilient and reparable in the wake of damage to its society and economy.

Let us take an extreme example that illustrates the invalidity of the organism metaphor. Assume that as the result of a nuclear war, half the population of either nation was killed. Assume also that the half that remained could gain access to the materials and skills needed to run a greatly truncated economy (i.e., one that yielded only a fraction of the prewar per capita income) and that these postwar survivors would do so even if there were damaging (but not intolerable) residual radiation and other obstacles. Some semblance of a society could, then, go on. Repair of damage to critical assets would be possible, and these particular repairs would yield large increases in postattack production capabilities. The point here is that a nation-state is one of only very few complex "organisms" that could recuperate from the destruction of half its bulk.

Numerous studies suggest that survivors can indeed "make do" even in the aftermath of highly destructive wars.* Postwar Japan, West Germany, and the Soviet Union offer spectacular examples that support assertions regarding the feasibility of relatively rapid recovery after great damage (damage comparable to that occurring in many of the scenarios we study). A thermonuclear war could present greater problems and uncertainties than destruction caused by conventional war,

---

* See Jack Hirshleifer, *Disaster and Recovery: A Historical Survey* (Santa Monica, Calif.: Rand Corp., 1963); G. Petty, *et al.*, *Economic Recovery Following Disaster; A Selected, Annotated Bibliography* (Santa Monica, Calif.: Rand Corp, 1977); Fred Charles Ikle, *The Social Impact of Bomb Destruction* (Norman, Okla.: University of Oklahoma Press, 1958); and Leon Goure, *The Siege of Leningrad* (Stanford, Calif.: Stanford University Press, 1962).

but depending on the details, probably could also be managed. Clearly, after suffering grievous losses, a country would never be the same again (i.e., there would be many long-term legacies of the war damage), but barring political breakup, the survivors would soon operate and function as a nationwide society. They would probably not "envy the dead."

As in almost all wars, postattack survival and recuperation would be greatly facilitated by appropriate prewar preparations. But even an unprepared society and economy should be able to recoup minimum necessities for survival and exhibit an important degree of social, economic, and psychological recuperation. There would, of course, be surprises and setbacks, but despite much discussion and analysis, we do not know of any reasonable scenario that precludes recovery in at least one and probably both of the warring countries.

Just as there are external sources that nations can turn to for crucial supplies in peacetime disasters, similar assistance would be available in many postwar circumstances. While the enemy might thwart access to these recovery resources or even obliterate many of them, it is not unreasonable to expect at least some to survive and be usable. If Japan and the so-called "newly industrializing countries" were relatively unscathed, they could play a major role in a U.S. recuperation effort. This is especially true for South Korea, Taiwan, Spain, Brazil, Mexico, Hong Kong, and Singapore. Furthermore, Australia, Canada, Italy, Sweden, Switzerland, Yugoslavia, and other medium-scale powers are unlikely to be totally devasted—at least in most scenarios; and if they are not, they could be very helpful in providing supplies in exchange for gold, credit, and normal trade. The importance of the U.S. as a consumer and producer would create strong incentives for other developed nations to promote U.S. recovery; doing so would clearly promote their own self-interest.

Accordingly, many analyses indicate that the United States or Soviet Union could survive and recover from even a fairly massive nuclear war.* While recovery from large attacks would probably not occur expeditiously (although it might), it would almost certainly occur eventually. And as long as there is a possibility of survival, we should have a serious and responsible interest in the subject and should make appro-

---

* See the catalog of such studies contained in Howard M. Berger, *A Critical Review of Studies of Survival and Recovery After a Large-Scale Nuclear Attack* (Marina Del Rey, Calif.: R&D Associates. 1978). Also see Jack C. Greene, *et al., Recovery From Nuclear Attack (and Research and Action Programs to Enhance Recovery Prospects)* (Washington, D.C.: International Center for Emergency Preparedness, 1979).

priate preparations, regardless of the unpleasant and terrifying nature of the task. Furthermore, most of the (more plausible) scenarios we study indicate that the problems are much less severe and the possibilities for recovery much greater than is usually envisaged.

One useful framework for thinking about a nation's potential for recovery after suffering massive war-related damage is to divide the nation into three parts: the "A, B, and C countries." The "A country" contains the highly urbanized areas with the majority of the nation's wealth and productive capacity. Most of the rest of the nation is located in the "B country" (the hinterland). The "C country" contains defense installations (e.g., strategic forces, logistics support facilities, command-and-control assets) which are likely to be high-priority targets in most nuclear wars.

A, B, and C countries can be thought of as more or less independent entities even as they interact and occasionally overlap with each other. For example, important military targets (C country) may be located in or near major cities (A country), so that it is difficult to demarcate the three countries precisely, but to a remarkable degree, the tripartite distinction is valid. The rural B country of the United States (and to a lesser extent that of the Soviet Union) could almost certainly survive separately from the A country; the urban A country could also survive if it had enough trade with B or with foreign areas; the military C country would do less well than either of its counterparts, since its role as an operating entity depends almost totally on the continued functioning of the A country. The B country in the United States is extensive, resilient, and quite capable (with even minimal preparations) of rather large productive efforts in the aftermath of a nuclear war. It would take a major military effort to inflict considerable damage on it, particularly if there had been advance warning allowing some time for evacuation, fallout and other population protection, and preparations for postattack recuperation.

In a country like the United States, the B country contains almost all towns of less than 25,000 people and all rural areas lacking important military installations. It probably contains more productive capacity than existed in the entire United States a quarter of a century ago. In a nuclear war, the B country would be damaged, but most of the knowledge, skills, and other capabilities for rapid recuperation would remain. Insofar as fallout is concerned, unless there were massive and deliberate (and probably impossible or militarily foolish) attempts to saturate all settled parts of the United States with nuclear-weapon effects, rather simple civil defense preparations (e.g., expedient fallout shelters) would ensure the survival of most of the population in the B country. (This would of course include most of the people from A and C countries who had been evacuated to B country.)

At a minimum, therefore, prewar preparations should be made to assist the survival and recuperation of B country. After that, we can turn our attention to protecting citizens in the A and C countries. Of the various measures to be taken, "crisis relocation" is the most important. When crisis relocation is feasible (and, as noted, it would be feasible in most plausible scenarios), its cost-effectiveness dominates that of all other measures to defend people from the effects of nuclear attack.*

The B country, in both the United States and the Soviet Union, could be prepared to receive evacuees from the A and C countries. By using spaces in existing buildings and by constructing other expedient shelters, most of the population could be reasonably well protected from fallout—and to some degree from the effects of blast and thermal radiation effects—on fairly short notice (a few hours to a few days). In some areas, particularly in winter, more time or more solidly built shelters would probably be needed.

Crisis Relocation and Crisis Bargaining

The major contention of this chapter is that crisis relocation and other civil defense programs are prudential measures for saving tens of millions of lives in the event of nuclear attack. A government might also order an evacuation of its cities to increase its bargaining leverage in a crisis. With an evacuation, the effects a nation's population might suffer in a nuclear war would be mitigated. With most of its population in places of relative safety, a nation's decision makers may feel more willing to play a stronger hand in the crisis bargaining. Nuclear threats made in combination with crisis relocation are more credible than the same threat of attack without evacuation. Despite the potential value of

---

* Despite many arguments to the contrary, evacuation during a crisis is not hard to do. Millions of people "evacuate" major metropolitan areas every holiday weekend. Evacuations in response to natural disasters occur often, expeditiously, and with little or no "panic." (See, for example, Walmer E. Strope, et al., "Importance of Preparatory Measures in Disaster Evacuations," Mass Emergencies 2 (1977, 1–17.) With regard to wartime evacuations, one can cite examples from British experience in World War II. At the start of the war, the British Government evacuated a million and half of its citizens from large cities in three days. Another two million evacuated without government supervision.

crisis relocation in enhancing the ability of the United States to withstand Soviet blackmail, this tactic would not contribute much (if anything) to the deterrence of a direct nuclear attack against the United States. Strategic forces-in-being provide this Type I Deterrence.

In a particularly tense situation, the Soviets could deliberately evacuate their A and C countries in order to put pressure on the United States or Western Europe to acquiesce to their demands. Eighty percent of the 100 million people in the 300 largest cities of the Soviet Union could be evacuated to their B country, leaving 20 percent in the cities to maintain all essential functions. (These key workers would have access to an extensive system of urban blast shelters and would be relocated to the B country later—but still prior to U.S. strikes on Soviet cities.) Under these conditions, if the Soviets were to strike first against U.S. strategic forces and were reasonably successful, a U.S. retaliatory blow would kill no more than 20 or 30 million Soviet citizens and probably considerably fewer. Thus the Soviet Union might suffer proportionate losses comparable to those it sustained in World II and those Imperial Russia suffered in the Napoleonic War and World War I. The threat of Soviet attack would be "not incredible."

In this scenario, what kind of moves could the United States take to convince the Soviets to back down? Generally speaking, the necessary response(s) would make clear that the United States recognized the gravity of the Soviet provocation, and also create an appreciable shift in the balance of power in favor of the United States. There would be a need to demonstrate resolve and to make it credible (or at least not incredible) that the resolve would be translated into an attack if the Soviets were not reasonable.

One such move would be to place U.S. strategic forces at Def Con ("Defense Condition") II or III. This would increase their alert status and combat readiness. As a consequence, any U.S. retaliation for further Soviet provocation would be large.

But just going on alert may not necessarily be very persuasive. The credibility of the threat to use nuclear weapons is more a function of one's ability to accept the retaliatory strike than to carry out the threat itself. A government unable to provide a large measure of protection to its people would be hard pressed to force one that was able to do so to back down. A counterevacuation would increase the credibility of subsequent U.S. ultimatums and offer a temporizing alternative to an immediate first strike or capitulation. While crisis relocation by both sides might increase the danger of war, it would also increase the probability that the crisis would be resolved quickly. The actual outcome of the bargaining would depend on further details for the scenario.

Evacuation and counterevacuation, however, can be risky. Counterevacuation might push the Soviets to strike preemptively rather than back down or compromise. At worst, what Thomas Schelling has called a "reciprocal fear of surprise attack" might be reinforced by mutual evacuation. In such a situation, each side would feel increasing pressure to strike first in order to beat the other side to the punch, even though both sides would prefer compromise to war.

In addition, the effect of an evacuation on the resolve of the American people and their European allies might be very different from the effect on U.S. decision makers. The level of tension and fear of war among the people may actually increase. Depending on the details of the crisis, the success of the evacuation, and the appearance of protective arrangements, the people are as likely as not going to be an influence for moderation, accommodation, or even appeasement. Alternatively, if the crisis remained unresolved, it is difficult to imagine a democratic nation meekly backing down and asking its people to return to the cities. (Even an authoritarian nation like the Soviet Union might have trouble manipulating its population under these circumstances.) Of course, during the most intense crises, the public will have little say.

Similarly, a greatly heightened fear of war in Western Europe might cause U.S. allies to put pressure on the United States to compromise in order to preserve peace at any price. Or they might accommodate the Soviet Union on their own via "preemptive surrender."

In short, difficulties and possible public and political reactions make the ordering of crisis relocation a momentous decision, and one whose consequences cannot be reliably predicted. Nevertheless, the United States should further develop crisis relocation as an option and be prepared to employ it in a way that maximizes the chance that crises will be resolved short of central war and without U.S. capitulation.

Is Civil Defense Both Ineffective and Too Effective?

Many critics assert that although civil defense measures would not work well enough to be worthwhile, they would still be effective enough (or *appear* to be sufficiently effective) to accelerate the "arms race" or make the outbreak of nuclear war more likely. Their argument is this: If the civilian population is a target, then civil defense programs will simply compel the Soviets to build more nuclear weapons to overwhelm these protective measures. Or equally important, if the Soviets fear that our civil defense preparations, by making our population less vulnerable to retaliation, increase the chances of our striking first in a crisis, they may feel it necessary to place their forces on a "hair-trigger" alert or to preempt.

I agree that some of these and similar problems might be raised by a large, crash, civil defense program, one, say, that was initiated at substantially more than $5 billion annually. But the kind of program recommended by the present administration (i.e., one costing *less* than $5 billion, over a *seven*-year period) will not greatly affect the strategic-arms competition or the risk of nuclear war.

For reasons previously discussed, neither the Soviet Union nor the United States is likely to procure and employ nuclear forces on the basis of how much of the enemy's population these weapons can destroy. Moreover, the MAD notion that deterrence will be weakened unless the total annihilation of the Soviet people is guaranteed, and conversely, unless we can promise the Soviets that a strike by them would kill every American, wildly exaggerates the willingness of either side to strike the other.

The argument that civil defense would be "too effective" flows from what might be called the "doomsday ethic." It is a little like sermons of a Puritan preacher, which threaten the flock with the damnation of hellfire and brimstone on the assumption that devout behavior will not result from a careful weighing of costs and benefits. Only the threat of unbelievably strong and inevitable punishment will regulate the behavior of the congregation. According to the doomsday ethic, you must not allow the slightest possibility or hope of surviving a nuclear war, or the slightest possibility that a clever military planner or a reckless political leader might find an opportunity for waging war while avoiding total annihilation.

My belief is contrary to that of the doomsday ethic. The United States and the Soviet Union could protect every one of their citizens with 100 percent reliability, and I still would expect both sides to be deterred in most cases. Civil defense measures are unlikely to cause significant changes in the general views national leaders hold regarding the consequences of nuclear war. Neither nation would be willing, under almost any plausible circumstances, to risk losing the buildings and facilities in their great cities—so valuable in economic terms and so rich in historical, sentimental, and cultural value. And neither nation could be certain that its protection and recuperation plans would work in practice. In addition, a country is not going to go to war lightly just because it could reduce fatalities from, say, 60 million to 20 million. Thus, civil defense is one means of avoiding the serious moral and political problems associated with mutual assured destruction while not diminishing mutual vulnerability to the extent that the deterrence of nuclear war is unduly jeopardized.

There are conceivable circumstances in which certain kinds of civil defense programs might tend to convert an especially tense crisis into a war. We would confront harsh choices. We might have to choose between the risk of immediate war and appeasement or surrender, with whatever that entails in terms of future risks. In such a situation, some civil defense measures (particularly urban evacuation) could well affect one side's calculations and increase the risk of war. As I argued in the previous section, evacuation could also *diminish* the danger of war.

In any event, it would be irresponsible not to recognize that such risks may have to be taken, and that having a credible ability to accept that risk may deter the Soviets from deliberately creating the very situation in which it would arise. The simple position that there is no alternative to peace (no matter how harsh) is a positive encouragement to the Soviets to create the situations in which the only alternative to appeasement is a great risk of war. Thus, it increases the likelihood of crises that can get out of control. An appropriate civil defense program might contribute to a relatively large reduction in the probability of war by reducing the frequency and intensity of Soviet-inspired crises.

It is important to understand that many of the actions we take to defend our national security and promote world order may increase the probability of war. In coping with the prospect of nuclear war, we should be willing to accept measures that increase the likelihood of war somewhat, but at the same time would greatly lessen the damage war might cause. A certain policy or program (like civil defense) on balance might be desirable if it increased the chance of war by, say, one in 10 million but also increased survival prospects by a factor of 1,000.

It is curious that those who are most pessimistic in their estimate of the probability and consequences of a nuclear war are generally to be found among the opponents of civil defense. If a person believes both that a thermonuclear war is possible and that it would be so horrible as to make life for the survivors more difficult than it has ever been in the whole of human history, then it would seem that that person should feel all the more obligated to advocate measures that could reduce casualties and ease the lot of the survivors, rather than merely abandon them to an untempered fate.

Conclusion

All of the above implies that if there is a reasonable possibility for the survival of society after a nuclear war, we have a moral obligation to prepare in advance such facilities as would help people to meet the emergencies they would have to face and as would improve their ability to improvise and organize. That there *is* help that can be given is indicated by every existing serious study. These studies are not complete and they are certainly not infallible, but they do give us enough ground for supposing that not all of the survivors of a nuclear war would meet objectively insuperable obstacles, especially if they were supported by a set of reasonable advance preparations for survival and recovery.

What do such preparations consist of? Among the most important are those designed to cope with immediate survival needs and to restore stocks on hand to last until production can begin again, and preparing schemes and facilities for distributing these stocks under the many different circumstances that could arise.

Civil defense is a good example of an effort in which meeting more than our first-priority objectives may be essential; as always there is a long list of things that we cannot safely do without. We need to eat, drink, keep warm, sleep, as well as breathe, and though human biology tells us which comes first, none can be ignored indefinitely. Similarly, there is no question that it is far more important to avoid war than to find ways of reducing its damage and of recovering from its effects. But since we can never be certain that we will succeed in preventing war, it is essential that we take moderate and prudent steps to minimize the disaster that such a failure implies. In the event of a war, civil defense measures could not only save millions of lives but could also prove critical to the continued survival in the world of Western ideals and institutions.

9

ARMS CONTROL

"Arms control is neither sin nor salvation. It is a way—along with diplomacy and military decisions—of managing Soviet-American competition." That perspective, offered in an article by Leslie Gelb,* seems to me to be exactly right. Arms control is an important part of an adequate defense program, but it certainly cannot substitute for one, or even be the main reliance of national security.

Yet many people argue that arms control is, in fact, salvation, or the only hope of getting there. Others argue it is a sin, that it is foolish and irresponsible to negotiate with a diabolical regime. They see no merit in putting the United States in a position where the Soviets can manipulate and take advantage of our negotiators and of U.S. public opinion in order to weaken U.S. defense preparedness. Since I tend to agree that this can be a problem, it is very important to insulate, at least to some degree, arms-control negotiations in particular (and defense preparations in general) from short-term, transitory political influences. For this reason, a long-term perspective is crucial to their success.

The immediacy of political pressures can, however, make a difference in how seriously an administration pursues arms control. No ad-

* "A Practical Way to Arms Control," *New York Times Magazine,* June 5, 1983, p. 33.

ministration should be prepared to ignore arms control per se; what is at issue is the kind of limitations to be negotiated (e.g., a nuclear freeze, a phased "build-down" in strategic nuclear weapons, deep reductions in ICBMs), as well as the priority of these negotiations in terms of competing foreign policy interests. The widespread call for a nuclear freeze has been useful in reminding President Reagan and his advisors that the American people are presently deeply concerned about the nuclear-arms competition, more so than they have been for some time. Similarly, the Pastoral Letter on War and Peace of the National Conference of Catholic Bishops was extremely effective in emphasizing—to the Left as well as the Right—the Church's concern, especially its belief that targeting nuclear weapons against civilians is inherently immoral.

This kind of short-term (or even "trendy") political influence on arms control is understandable as well as unavoidable, but negotiating away any long-term national security advantages or compromising any U.S. defense interests for temporary political expedience seems to me to be unacceptable. Defense programs require sustained—not occasional—public support. Partly, this can be achieved by effective government officials making their case in a responsible fashion. If, for example, they can persuasively argue the need for a new weapon system and explain that most such systems have a lead time of a decade or more— this in and of itself should counter some of the pressures for short-sighted accommodations.

The kind of international political considerations that I believe should not be linked to arms-control negotiations include issues of economics, trade, human rights, and so on. While all are to some degree relevant to international relations, they are not usually matters of long-term significance to diplomacy, foreign policy, or national security— narrowly defined—and these latter are the substance of arms control. Neither short-term political "quick fixes" nor purely military solutions are likely to be adequate in the long run.

Nevertheless, current international political problems can get in the way of arms negotiations. Neither side is eager to pursue agreements in a tense international atmosphere, but they may have to. It may be even more dangerous to wait until all outstanding tensions are resolved before pursuing less ambitious goals that are of mutual interest. Indeed, arms control could reduce the probability of a war-causing type of event actually causing a war. A limited U.S.-Soviet consensus on at least some issues should be possible at almost any time (for example, the

Limited Test Ban Treaty was negotiated soon after the Cuban missile crisis).

Neither the United States nor the Soviet Union wants a war that would annihilate us both (although the Soviets could manipulate the common fear of gaining a one-sided advantage); neither side wants the cost of the arms competition to become more onerous; neither side wants to permit lax operational practices for nuclear forces, which might serve as the proximate cause of a nuclear war. Common objectives do exist, even as each side seeks to improve its relative position. Many "restraints" that can be put into treaties or agreements are not "concessions" but recognition of the adverse consequences each side would suffer if it acted otherwise.

Yet arms control is also another element in the political-military bargaining process between the U.S. and the Soviet Union. In fact, the arms-control arena itself becomes a substitute for direct military competition, where each side attempts to emerge victorious on the propaganda front while trying to gain every substantive advantage for any possible real war.

While there is a widespread tendency to equate "arms control" with formal, bilateral agreements, there are many measures that do not have to be made explicit; they are followed simply because they are either unilaterally useful or accepted as custom—the unquestioned code of behavior. One such custom involves intelligence operations; the unwritten rule that one does not commit unprovoked violence against the other side. To some extent, there is tolerance of violence carried out on one's own dissidents, defectors, or schismatics (e.g., the murder of Trotsky at Stalin's behest), but the attempt to use violence in other circumstances (e.g., U.S. plot to assassinate Fidel Castro) is unacceptable.

Another implicit mode of conduct involves conversations held with the president in the White House: "No lying in the Oval Office." The Soviets breached that rule at least once—they lied to President Kennedy about missiles in Cuba (Foreign Minister Gromyko saying they were strictly "defensive" and could never constitute a threat to the U.S.).* The "sense" of knowing which transgressions are permitted

---

* More recently, President Brezhnev, in a personal letter, "lied" to President Carter about the Soviet invasion of Afghanistan. In an interview conducted shortly after he received the letter, Carter described its contents in the following terms:

Brezhnev claimed that he had been invited by the Afghan Government to come in and protect Afghanistan from some outside third-nation threat. This was obviously false because the person he claimed invited him in, President Amin, was murdered or assas-

and which are not is an important part of arms control. It is also important that the U.S. and the Soviet Union maintain trustworthy channels of communications.

An informal measure more specific to arms control was initiated in late 1962 when the Americans let the Soviets know how to "accident-proof" their nuclear weapons, thereby reducing the risk of inadvertent war. The critical information was relayed to the Soviet Union in two ways. First, in a public speech delivered by a high-ranking Defense Department official that described U.S. efforts to develop command-and-control systems to prevent unauthorized or accidental use of nuclear weapons. Second, following the speech, several U.S. arms-control specialists met with Soviet scientists and academics to focus their attention on the importance of the speech, and to provide further details on how they could similarly improve the safety of their nuclear weapons. The Soviets apparently got the message and developed the ability to place their nuclear forces on alert in a safe manner (a development not all defense analysts welcomed).

This quite important arms-control measure was never formalized or explicitly acknowledged by any kind of official agreement. (The late Donald Brennan of the Hudson Institute participated in some of the unofficial discussions with the Soviets.)

Arms-control efforts seek to lessen the dangers of nuclear war. The objectives are to diminish the likelihood of war and to reduce the damage of war should deterrence fail. Any steps that contribute to those objectives (better command-and-control systems, more survivable military forces, new "fail-safe" techniques) can be considered "arms control," even though they may be adopted unilaterally and confer unilateral military advantages in wartime. Sometimes a "safer" world can be achieved through arms buildups, sometimes through arms reductions; it cannot be achieved by disinventing nuclear weapons. If Britain and France had in fact engaged in an arms race with Hitler in the 1930s,

---

sinated after the Soviets pulled their coup. He also claimed that they would remove their forces from Afghanistan as soon as the situation should be stabilized and the outside threat to Afghanistan was eliminated. So that was the tone of his message to me, which, as I say, was completely inadequate and completely misleading.

Question by interviewer: Well, he's lying, isn't he, Mr. President?

Carter: He is not telling the facts accurately, that's correct.

(Quoted in "Transcript of President's Interview on Soviet Reply," *New York Times,* January 1, 1980, p. 1.)

World War II might have been averted. Under the present circumstances, the best we can do to enhance "safety" is to come out strongly on the side of no first use, decrease the influence and status associated with nuclear weapons to discourage any further proliferation and the resort to nuclear threats for foreign policy purposes, and create well-planned and well-understood limitations on their use if deterrence breaks down. (These topics are taken up in the next chapter.)

Many scholars of military affairs have argued that arms control and military strategy are basically the same. One can certainly define the two to make this true, but it is more useful to distinguish rather sharply between them, even more sharply in theory than would be desirable or possible in practice. A "strategist" seeks unilateral advantage for his country, side, or cause. An "arms controller" seeks increased security and diminished risks and costs for all concerned. A strategist is basically competitive in his outlook; he emphasizes the "zero-sum" aspects of the situation, i.e., his side's gains are the other side's losses. An arms controller is basically noncompetitive, looking instead at mutual and shared interests.

Yet even a relatively one-sided strategist often pursues policies that the arms controller would applaud and that are dependent not on negotiated arms-control arrangements but on self-interest. Similarly, in many cases one might be able to use arms-control negotiations in a strategic fashion by "tricking" the other side into seeing significant mutual gains when the gain is in fact fairly one-sided. "Strategists" and "arms controllers," therefore, are rarely totally "pure" groups.

In some (very limited) ways, the doctrine of mutual assured destruction (MAD) is the quintessential form of the arms-control concept—i.e., a fairly even distribution of possibilities for mutual gain and avoidance of mutual loss. Neither side wants war to break out. If it does, both sides are destroyed, and both sides understand this without any complexities and uncertainties. Some arms controllers desire to make this "fact" as stark as possible. But the very concept of deliberately subjecting nations to an all-or-nothing risk of nuclear war, even if inadvertent, is almost the exact opposite of what arms control should be about. Further, since MAD (or any "balance of terror") can be used as a cover for one-sided aggressions, a MAD posture could be the essential basis of an adversary's campaign to manipulate the fear of war to impose diplomatic defeats on his opponent.

Pressures to reduce the hazards posed by nuclear weapons will result in both responsible and irresponsible antinuclear and "peace" movements. The Soviets have sought to exploit these movements and the fear of war to disarm (psychologically as well as militarily) the U.S. and its NATO allies and otherwise manipulate the internal political movements in the West to their own advantage. Both sides have tried to develop persuasive arguments as to why the other side should make political concessions or accommodations to reduce tensions as part of their bilateral arms-control negotiations.

Hence, while MAD can be viewed from one perspective as a strategist's nightmare and an arms controller's dream, from another perspective this association can be weakened (or even reversed). A more effective and realistic arms-control objective than "peace at any price" (the aim of a posture restricted to MAD—i.e., symmetrical U.S. and Soviet Type I Deterrents) would be "multistable" (Type I plus Type II) deterrence, that is, a posture designed to prevent a first strike, provocations, and coercion employing nuclear threats.

From the viewpoint of global prospects in general and NATO prospects in particular, we prefer the asymmetrical approach of the strategist (with the asymmetry on the side of NATO) to the more balanced view of the arms controller. This corresponds to Churchill's point in the interwar period that those British who thought it unfair to have the Germans weaker than the French as a result of the Versailles Treaty were misguided. Fairness was not the issue; if the Germans ever became stronger than the French there would probably be war; as long as they were weaker, there wouldn't. Those who preferred peace should prefer to support the military imbalance created by the Versailles Treaty. Churchill's basic point (and ours) is simply that the evaluations of the arms controller and the strategist are very much a question of perspective. To some degree, everyone should be both a strategist and an arms controller.

To illustrate the possibility of a common interest even between two enemies who would like to fight a duel to the death, consider a situation in which two combatants intend to duel in a warehouse filled with high explosives, using blowtorches as weapons. It is conceivable that they would agree to keep the lights on: they want only one combatant to survive, but they want to ensure the possibility that one will. Even if one side would gain more than the other from having the lights on (because of greater visual acuity, and so on), the other side would agree simply because no other option would give it an appreciable, even if asymmetric, probability of surviving. Agreement, however, would be much more difficult if it were not simply a question of lights on or off, but of what kind of lights, where should they be placed, what time should they be turned on, who should turn them on, and so forth.

An analogous situation points up an important characteristic of arms control. Very often the only agreements that are possible are those that involve some kind of conspicuous natural demarcation. Usually, that means maintaining the existing military balance and permitting the completion of planned weapon programs, while placing limits on further developments. For example, the SALT II agreement allowed each side to fulfill its current arms agenda, but neither side could start many other new programs.

One important issue is the degree to which each side estimates the common danger of the "arms race" as opposed to the danger from each other. During most of the postwar arms talks, many American negotiators seem to have been more afraid of the "arms race" than of the Soviets, i.e., their fear of nuclear weapons per se seemed greater than their fear of Soviet use of them or lesser Soviet acts of aggression. This particular anxiety often has led the United States to adopt imprudent negotiating tactics in its arms-control dealings with the Soviets. These include:

1. Be in a rush for an agreement
2. Always be willing (or under pressure) to:
    A. Replace a proposal considered "unacceptable" by the Soviets with a more favorable one
    B. Offer unilateral initiatives
    C. "Split the difference"
    D. Present "sincere" proposals
    E. Show flexibility on "agenda" issues
3. Make (or let) the Soviets believe that we:
    A. Feel that our common humanity should be the dominant consideration in the negotiations
    B. Are more afraid of the "arms race" than of the Soviets
    C. Always "mirror-image" (i.e., view Soviet attitudes, policies, and positions as similar—if not identical—to our own)
    D. Believe in nuclear "overkill" and/or that strategic parity implies stability
    E. Prize negotiations for their own sake
    F. Prize agreements for their own sake
4. Actually base negotiating positions on any proposition in "3"

This situation was by no means reciprocal. Soviet negotiators feared the potential use of American military power much more than the arms competition itself.

Part of this asymmetry is rooted in national character: Americans tend to believe that an accidental event (such as a defective microchip, a circuit failing, or a button pressed inadvertently) could cause a nuclear war and thereby change history in a fundamental way. Soviets tend to think otherwise. It is not that they are unaware of the potential for trouble from these causes, they clearly are. But they do not take them that seriously, and they recognize there is not much they can do about them in any case. Soviets trained in dialectical materialism still believe that major events reflect fundamental historical forces. They think it unlikely that a dialectic of history will be influenced by trivia, that basic political, economic, and social conflicts between nations are more important causes of war than accidents or inadvertencies. As a result, the position of strategist is much more natural for Soviets than that of arms controller.

Although some Soviets hold views very similar to those we ascribe to Americans, and vice versa, the above is a valid generalization, especially for the Politburo members and other senior Soviet military and civilian officials. Their basic outlook is essentially that of a traditional strategist; in public, however, they increasingly use the language of the more "balanced" arms controller to exploit the antinuclear movement in support of Soviet political and military objectives worldwide.

This basically "strategist" orientation is also true of much of the so-called political Right in the United States and Western Europe. As far as the New Left is concerned, its orientation is much more toward arms control. We attribute this to a basic lack of realism or to an illusioned idealism, or sometimes to an innate propensity of many "peace" groups to be sympathetic to communism, if not outrightly partisan.

The viewpoint of the arms controller comes quite naturally to those who find the current military balance between the United States and the Soviet Union to be reasonably acceptable. If one has a strong desire to change the world (through either war, revolution, ideology, or crusades), emphasis on preserving the status quo is a less attractive alternative. However, this does not prevent even revolutionary regimes from using arms-control negotiations and attitudes when appropriate to their cause. (In fact, they are delighted to make unserious and unrealistic but popular proposals, such as many of the Soviet proposals put forth over the past thirty years.) Yet, even if one side is revolutionary to the extreme, there can still be arms-control proposals that serve mutual interests, such as many traditionally honored "rules of warfare."

While arms control may often involve a degree of disarmament, it is not synonymous with letting down one's guard. Some civilian defense analysts and military planners have trouble accepting the need for any arms negotiations; to them it is the equivalent of selling out. But most responsible people involved with national-security matters understand the urgency of controlling the arms competition in a constructive, long-term fashion. Stable arrangements for the long-range future can be the best way to reduce the danger of U.S.-Soviet confrontation and decrease uncertainty in military planning, even if large military establishments on both sides remain. Even successful arms control probably would not lead to smaller military establishments in the foreseeable future (the history of the SALT negotiations supports this assertion). This is not necessarily a cause for excessive concern. The sheer size of the nuclear arsenals on both sides has little to do with the probability that war will break out. Indeed, large forces contribute to a more stable balance of terror, and to an ability to make (relatively minor) concessions and to accept the breach of (relatively minor) violations of whatever agreements exist.

One central problem of arms control is how to deter, punish, and correct for Soviet violations of arms-control agreements. Many (either tacitly or explicitly) offer the "outrage" of public opinion as a check on Soviet violations. The trouble is, indignation is unlikely to be very effective. The violation may be perceived as ambiguous and therefore a candidate for rationalizations by the offender, his sympathizers, or weak-willed opponents. The limited practical consequences of international attempts to ostracize the Soviet Union for its invasion of Afghanistan are also instructive in this respect.

A second and considerably more compelling deterrent would be the threat of a deterioration in relations between the United States and the USSR, and all that would entail—greater hostility, heightened distrust, strong impediments to additional agreements, and an accelerated technological and material arms competition. As I argued in chapter 7, an effective mobilization base would make the last of these "penalties" very credible to any Soviet leaders contemplating the costs and benefits of violating an arms treaty.

It is important to seize the moment when arms-control negotiations might yield the best results. I believe that the mid-1980s is probably a very good time for this. For one thing, the Soviets will be facing huge economic problems during the rest of the decade. Under the circumstances, they could probably still exceed their planned production of nuclear arms (and other military products), but they would prefer not to —both because of the costs to their own economy and society and because they recognize they could not win a genuine arms race with the United States. The Soviets also realize that trends within the United States are against them. It now appears that President Reagan may win a second term and, with his victory, be able to sustain the momentum behind the current U.S. arms buildup. In addition, the present and anticipated growth in the recovering economy could support an even greater rearmament program. After the initial deployments of intermediate-range nuclear missiles, MX ICBMs, B-1B bombers, air-launched cruise missiles, and Trident submarines, the U.S. could threaten to produce additional units in an almost assembly-line fashion. If the Soviets want to head off this possibility, it is a lot smarter for them to reach an arms agreement with the United States sooner rather than later, when such a further U.S. buildup may be considered necessary.

The necessity of taking arms-control measures seriously—and of punishing violations—is a matter of self-interest. They can help make a war less likely and if it does occur, less costly in human suffering and material loss. Some kinds of dangers can be the subject of direct negotiations, while others lend themselves only to tacit understandings or unilateral initiatives.

The primary purpose of arms negotiations should not be to reduce defense budgets. In fact, the more carefully we limit nuclear weapons, the greater conventional arms spending and expenditures for active and passive defenses are likely to be. Even agreements that might include a build-down in nuclear forces would not automatically free up funds presently committed to defense programs. Arms-control measures are only one part of a realistic defense posture. A nation's defense establishment exists to fight a war in the event one has to be fought, even as negotiations proceed to make that contingency less likely. This is no less true of the nuclear era than of any other time in history.

An emotional commitment to arms control must be coupled with a practical commitment to national security, to making our nation—and the world—as safe as possible. We are trying to preserve peace through military as well as diplomatic means—not only because we want to survive, but because we want to have justice and prosperity, and they are part of what we mean by peace. On the other hand, justice and prosperity may require accepting grave risks. They may also require the use of violence at many levels. The objective of arms control is to help limit that use as much as possible.

Some Comments on Multipolarity and Stability

Whether or not adequate arms-control agreements are negotiated, it seems likely that over the next twenty to thirty years the world will become intrinsically safer than it is now (though in no sense as safe as one would like). By the end of this century (or soon afterward), the present international predominance of the two superpowers may have been so eroded through economic, technological, and political changes that the term will be abandoned, and the concept of "Great Power" status will be revived. A constellation of Great Powers, most of them armed with nuclear weapons, is likely to emerge more or less naturally. These Great Powers will take places alongside, or at least not far behind, the United States and the Soviet Union.

The result of this change in the international system will probably be greater stability in some ways, less in others. However, it is almost certain that in this multipolar world, most remaining MAD policies will be superseded by nuclear-weapon doctrines that place greater emphasis on restraint in the use of nuclear threats, limitations in the conduct of nuclear war, and defense against nuclear attack. Such a course of events would not reduce the desirability of many kinds of arms-control measures, but it could, by itself, lessen the risk of great international catastrophe. Indeed, it is likely to make a greater contribution to the control of nuclear arms than most or all arms-control agreements now under discussion.

Around the year 2000, there may be seven nations that are economically and technologically very advanced. Their trillion-dollar gross national products will set them apart from the rest of the world. The following table lists these powers and provides surprise-free projections of each nation's gross national product, population, GNP per capita, and military budgets. These are reasonable forecasts based mostly on a continuation of current trends; they do not factor in dramatic and abrupt national or international changes that might occur. The other five countries listed in the chart (plus several others not named), will be powers of a lesser order, but their substantial national wealth and technological capabilities should make them an important factor in their regions, and even on the larger world scene.

This table is relatively optimistic about the economic future of the world. It also assumes a world that is moderately, but not intensely, hostile and well armed. (The high levels of arms suggested in the table

WORLD POWERS OF 2000 (1980 DOLLARS)

Country	GNP* (Trillions)	Population (Millions)	GNP/Cap.* (Thousands)	Military Budget (Billions)	% GNP
United States	4½	260	17	200–500	4–11
Japan	3¼	130	25	100–400	3–12
Soviet Union	2½	300	8	250–500	10–20
China	1½	1,300	1⅕	100–300	7–20
Germany	1¼	60	21	30–100	2–8
France	1¼	50	25	30–100	2–8
Brazil	1	200	5	30–100	3–10
India	½	1,000	½	25–50	5–10
U.K.	½	60	8	20–40	4–8
Mexico	½	100	5	15–20	3–4
Italy	½	60	8	15–25	3–5
South Korea	¼	50	5	20–35	8–14
Rest of World	6½	2,400	2¾	150–650	2–10
Total	24	6000	4	1000–3000	4–12

* We use integers and fractions in the above columns to emphasize the roughness of the estimates and to make some fine distinctions without giving an impression of excessive preciseness. All dollar figures are in terms of 1980 dollars.

are pessimistic, but could happen.) Most of the seven "trillionaires," and possibly all of them, will have Great Power status by 2000 or soon afterward. The military capabilities of the two present superpowers may exceed those of the new Great Powers, but at least one of the powers—Japan—clearly will have the basic technological and economic wherewithal to compete with the United States and the Soviet Union.* The U.S. and the Soviet Union might still dominate world affairs in 2000, and, in effect, still have the status of superpowers—but they will no longer be as disproportionately dominant. Dramatic changes in either NATO or the Warsaw Pact are also possible and indeed likely, as is the

* Of course, for Japan to acquire nuclear weapons, current domestic opposition in that country to such a development would need to be overcome. For a discussion of how Japan might "go nuclear," see Lewis A. Dunn and Herman Kahn, *Trends in Nuclear Proliferation, 1975–1995* (Croton-on-Hudson, N.Y.: Hudson Institute, 1976), pp. 51–55.

creation of other regional or world military organizations. These could also affect the status of superpowers as well as Great Powers.

The transition to a multipolar world could involve significant regional risks and instabilities (e.g., those arising from the nuclear armament of Japan and West Germany). If these became great enough, they might even prevent the transition from being successfully negotiated. However, once it was achieved, a world of multiple nuclear-armed Great Powers generally would be less prone to cataclysmic nuclear war, and even nuclear crises, than is the current international system.

In many ways, a future multipolar world would be similar to the balance-of-power system which obtained in Europe from 1815 to 1914. It is fair to say that from 1871 until its failure in 1914 through a series of coincidences, deterrence via the balance of power kept the peace. During this period the major nations of Europe remained at peace despite the fact that large segments of each nation desired war on numerous occasions, that there were bitter national antagonisms, and that the consequences of war were not only nowhere near as serious as today, but were regarded as far less serious than they actually were. A nuclear-armed multipolar world would partake of the aspects of the balance-of-power system that are conducive to international equilibrium, while avoiding some of the fatal instabilities.

The innate structure of multipolarity would lend itself to more stable arrangements than those achieved through arms-control treaties alone. When only two predominant nations compete for power and influence in the international system, the possibility of a direct confrontation between the two always exists. Every local conflict occurs within the context of this worldwide competition. A bipolar world almost inevitably results in recurrent crises between the two dominant powers. Moreover, structural reasons for reconciling their differences are lacking.

The situation is not much more stable when a third power is added: two of the nations are likely to get together, on the basis of either ideological compatibility or sheer pragmatism, to oppose (or destroy) the third. A four-power system splits too easily into two hostile coalitions, with one of the two powers on each side generally dominating the policies and actions of the alliance.

In a five-power world, one of the powers is likely to be, and seeks to be, a balancing power, throwing its weight to the weaker or less aggressive side as the balance of power or the source of political-military disequilibrium shifts. Its basic interest is in preserving both the peace and the balance of power. (This is the role that Great Britain historically played vis-a-vis the Great Power politics of continental Europe.) By shifting alliances, the fifth (or "swing") power can deter two stronger powers from taking advantage of any temporary military superiority, because even two stronger powers would be unlikely to try to defeat three. In turn, conflicts between any two powers among the five would be disadvantageous to both the victor and the vanquished, since the remaining—neutral—powers could only benefit from the decline in the strength of the two belligerents that would result from a war. Whatever damage was suffered in the war would be compounded by the new risks to be run in the postwar world. This prospect is likely to make any power think twice before precipitating a crisis or letting a crisis get out of control.

Other combinations and permutations are, of course, possible. Indeed, in the seven-power world forecast here, the coalescence of successful aggressive alliances among the Great Powers would be more difficult, and the possibilities for one or more balancing powers to emerge would be greater.

The emergence of a multipolar world would probably give a strong impetus to the adoption of more sensible nuclear-weapon postures by the U.S. and other countries. For example, a multiplicity of both great and small nuclear powers would increase the need for strategic defenses. Most of the Great Powers would not wish to be starkly (and unnecessarily) vulnerable to an attack by a two-bit nuclear power. If the Great Powers failed to develop active and passive defenses to complement their offensive forces, other countries could buy relatively simple nuclear-weapon systems, and using relatively simple (though perhaps dangerous) techniques for making these systems invulnerable, could then also claim to be Great Powers. Nuclear threats and nuclear deterrence between a great and a lesser nuclear power would then become two-way streets.

On the other hand, if the Great Powers did possess elaborate strategic defenses capable of protecting them against limited attacks, the smaller powers could not really challenge them. (Although the smaller powers would be able to acquire long-range missiles and bomber aircraft with which to deliver their nuclear weapons, they could not field the more expensive and more complicated systems needed to defend against nuclear attack.) In these circumstances, the development of strategic defenses would act to dampen the incentives for further nuclear proliferation, by emphasizing that big and small powers are not equal. A multipolar world in which the Great Powers also had strategic defenses would not be like the Wild West, where anyone with a six-gun could "play the game."

In a world in which all the Great Powers had to be careful about where an attack might come from, strategic offensive forces would have to be less accident-prone and more capable of waging controlled nuclear war. Planning and preparations for launch on warning would be more difficult to carry out because of the greater uncertainty regarding the identity of the attacker. The aim of limiting the destruction caused by a nuclear war (especially an inadvertent war)—a traditional objective of arms control that many people often forget—would become more important. Both belligerents in a nuclear war would have an additional incentive to keep the conflict controlled: the aforementioned postwar danger of falling victim to "laughing third parties" (Great Powers who were neutral during the war).

In sum, the emergence of a multipolar international system, apart from future progress in arms-control negotiations, could serve to make the world safer from the threat of nuclear war. Deliberate policies for achieving this same goal and reinforcing the stability inherent in the multipolar world are discussed in the next chapter.

PART FOUR

Making the World Safer

10

Seizing the Moral, Political, and Strategic High Ground

Four decades into the nuclear era, nuclear weapons seem to have taken on a new and immoral life of their own. They are alternatively cited as the harbingers of an almost certain Armageddon or as the last stop on the way to the redemption of mankind. There is a constituency for every antinuclear position, and a coalition of concerned doctors, lawyers, mothers, actors, and so on, for every antinuclear rally. But the real issue is how to achieve national security and international order within a morally and politically acceptable framework.

Three Popular (but Flawed) Antinuclear Positions

Under the circumstances, that becomes a very difficult task. Probably as much as any other single book, Jonathan Schell's *The Fate of the Earth* raised the antinuclear consciousness to the point where anything short of the elimination of all nuclear weapons (and all conventional forces) becomes morally and politically unacceptable. However, one of his fundamental arguments—that all values and interests must be subordinated to avoiding nuclear war—is impractical, illusionary, and dangerous.

If we take the position that Schell takes, that nuclear war clearly threatens the existence of humanity, that nuclear war has some significant probability of ending all human history, and that as far as humans are concerned "peace" has to be the overwhelming value and that there are no other values that in any way can compete with this, then we can get into quite a lot of trouble. Focusing our attention solely on "peace" (defined simply as the absence of nuclear war) can lead to weakness that creates opportunities for smaller scale (but nonetheless dangerous) aggression that threatens peace. There are two great defects in Schell's position: one of them is that you simply cannot afford to make any single value overwhelming. One must assign some kind of finiteness to any value because society cannot let one priority dominate everything it does. The second and more important issue is that making one value infinite still does not give a nation any direction—it does not tell government officials how best to preserve this value.

In order to establish the transcendental value of avoiding nuclear war at any cost, Schell describes, in graphic (and often exaggerated) detail, the incredibly horrible consequences of a nuclear war. The emotional impact of the book is apparently so powerful that very few readers (and very few reviewers) ever notice the book's substantive inadequacies and inaccuracies—e.g., the author's highly selective and tendentious use of evidence, or the distortions that necessarily follow from exaggerated assumptions about the dangers and risks of a nuclear war.*

And yet, if it is necessary to overstate the case in order to focus public attention on the nuclear threat, then Schell and the authors of many other recent antinuclear books may be serving a useful purpose. Nuclear weapons exist, they will not go away, and one day they might even be used. What happens then, however, is an issue that none of the volumes are willing to deal with in any useful or reasonable way. Schell's utopian solution envisions some kind of world government where sovereign nations cede their political authority and military power to an unclearly defined global order. How we get to there from here is one of the many "awesome, urgent tasks"—"the political work of our age"—that Schell says he has "left to others."

The nonsolution that has captured the greatest popular attention is the freeze. According to most advocates of this impressively widespread movement, both the U.S. and the Soviet Union should agree to halt any further testing, production, or deployment of nuclear weapons.

---

* These criticisms are elaborated in my review of The Fate of the Earth. See "Refusing to Think About the Unthinkable," Fortune, June 28, 1982, pp. 113–16.

In a vision that is almost as romantic and simplistic as Schell's, they believe a freeze would automatically end what they fear is an "inevitably spiraling arms race." Perhaps. But while it may prevent further growth in the U.S. and Soviet nuclear-weapon stockpiles, it would also prevent unilateral measures by the U.S. (or the Soviet Union) to lessen the dangers of nuclear war breaking out (e.g., through the deployment of more survivable nuclear forces) or to make nuclear war less horrible (e.g., through the development of lower yield, more accurate warheads). Like any across-the-board limitations (e.g., rent control and wage-and-price controls), the freeze is an indiscriminate restraint that is only a partial remedy at best and creates adverse side effects.

Further, a freeze would codify the current Soviet edge in nuclear forces (just as SALT was a formal acknowledgment by the U.S. that it accepted Soviet achievement of strategic parity and recognized the Soviet Union as a superpower). Granting the Soviets this edge now could have deleterious political implications for the U.S. and its allies.

Because there is popular support for a freeze in the U.S. and not in the Soviet Union, the freeze movement imposes asymmetric pressure on U.S. negotiators to reach ill-conceived agreements at the Geneva talks. And assuming a satisfactory bilateral freeze could be negotiated, what guarantees are there that the Soviets would in fact honor it? There is some evidence that they may have violated certain provisions of the SALT agreements (agreements with a relatively limited arms-control ambit), and even with relatively sophisticated verification techniques, human ingenuity can always devise ever more sophisticated ways around limitations.

Nor would a nuclear freeze preclude nonnuclear weapon system breakthroughs—potentially devastating developments that could be in the making even as a freeze is put into place, to be exploited against an opponent at the first signs of tension. A freeze would not have helped lessen any dangers at all if a crisis situation arose to cancel it—no arms reduction talks would have taken place, no further safeguards (political or technological) would have been worked out; in short, no meaningful arms control would have been achieved. The stagnation fostered by a freeze could only bring about a false sense of security that would be exposed as soon as the popular pressure subsided and a hostile confrontation seemed imminent.

The extreme wing of the freeze advocates goes so far as to advocate unilateral reductions in nuclear armaments, either regardless of the consequences or hopeful that such a show of good faith and honorable intentions would shame the other side into a similar position. Many proponents of unilateral disarmament believe in the so-called "demonstration effect," i.e., if the United States refrains from deploying the MX ICBM or Pershing II missiles, or other new nuclear weapon systems, the Soviets will exercise similar restraint. Unfortunately, the history of U.S. unilateral restraint in its nuclear-weapons programs is a sorry one, best characterized by former Secretary of Defense Harold Brown: "When we build, they build—when we stop, they continue to build."*

In many ways, the Green Movement in West Germany is the European counterpart of the freeze campaign in this country. (The label "Greens" stems from their initial concern with environmental issues.) The reasons for the strength of the movement in the Federal Republic are not hard to understand—most of the members are fairly young, postwar children of affluence. They have no experience with economic hardship or hostile geopolitical realities. Their personal satisfaction does not derive from material progress (they are all overprivileged), but from participation in "meaningful" social causes. In some ways they are the most protected, naive, and illusion-prone young people in the West today.

Add to this the fact that West Germany today is without nuclear weapons and therefore without the means of self-defense against Soviet nuclear coercion or attack. Moreover, the world's largest concentrations of conventional military forces face each other on either side of its border. Indeed, most scenarios for a serious confrontation or war between the U.S. and the Soviet Union evolve out of the potential for conflict between these two forces. In most situations, therefore, West Germany would be the battlefield (at least initially) in any war between NATO and the Warsaw Pact. As the "heartland" of Europe, the richest European member of NATO, and a nation with a border contiguous to the Warsaw Pact, it would also be the most likely prize. (One can envision many conflict scenarios that end with a de facto division or neutralization of West Germany.) Quite understandably, then, West Germans feel most at risk and have very little interest in pursuing confrontational policies. The Greens' call for West German neutrality (or reunification with East Germany) and the elimination of nuclear weapons from German (and European) soil is understandable within this context.

---

* For a discussion of the U.S.-Soviet "Arms Race," see my *U.S. Strategic Security in the 1980s*, HI-3212-P (Croton-on-Hudson, N.Y.: Hudson Institute, 1980), pp. 11–17.

The morality of any of the above positions—Schell's vision of a nonnuclear utopia, support for a nuclear freeze, opposition to European-based nuclear weapons and the existing European military alliances—is self-evident to their adherents; it escapes others entirely. I, for example, fail to recognize any redeeming higher value in a stance that places an entire nation at any greater risk than necessary. Flaunting one's vulnerability (which is essentially what a freeze would do) is neither morally nor militarily useful. Politically, it is primitive. A much more responsible posture would be to do as we are and have been doing —namely, to pursue bilateral arms reductions seriously (e.g., the ongoing Strategic Arms Reduction Talks [START] and the Intermediate-Range Nuclear Force [INF] Negotiations), while continuing to improve our national defense through a rearmament program. The principal moral obligation of a government in the nuclear age is to make every effort to enforce deterrence or, should deterrence fail, to limit as much as possible the damage to its citizens and its economy and to enhance the prospects for postwar recovery.

"Nuclear Morality" and the Pastoral Letter on War and Peace

Taking somewhat of a middle ground between the nonnuclear extremists and the more moderate antinuclear realists are the American bishops of the Roman Catholic Church. Their Pastoral Letter (drafted by the Committee on War and Peace of the National Conference of Catholic Bishops) attempts to provide spiritual guidance on how to reconcile the existence of nuclear weapons with traditional Church doctrine—and concludes that such a conciliation may not, in fact, be possible.

The bishops have attempted to apply the criteria for a just war to nuclear war. They conclude (with insufficient supporting evidence and analysis) that a just nuclear war is practically impossible because of the "overwhelming probability that a nuclear exchange would have no limits" and would therefore violate two of the major principles of the doctrine, namely, "discrimination" and "proportionality." By proscribing first use of nuclear weapons and the deliberate targeting of civilians, the Pastoral Letter also (correctly) condemns any attempt to gain security "on the cheap."

Contrary to the bishops' assessment, I would argue that the nuclear age has not rendered the doctrine of a just war obsolete. While it is true that no war can be reliably limited, it is not at all certain that all nuclear wars will escalate. An attempt to fight a limited nuclear war may be the least desperate choice in a future U.S.-Soviet conflict. Moreover, it has become more imperative than ever to meet the highest and most stringent criteria of justification for waging war, and—equally important—to strengthen both the effectiveness and the morality of deterrence.

In this regard, the Church makes an important contribution to the nuclear debate by endorsing no first use and by stressing the prohibition against the targeting of civilians and their property. I began advocating no first use about twenty years ago, and have continued to support it ever since. But the Pastoral Letter stretches no first use to a practically no-use-at-all position (basically nuclear pacifism) that I believe is unnecessarily dangerous and, for those of us who are not religious or philosophical pacifists, probably immoral. The letter explicitly calls for no first use but implicitly goes beyond that position to a no-use posture. It proscribes attacks against cities; it rules out attacks against many military installations because the collateral damage, even in limited nuclear wars, would be "disproportionate"; and it condemns counterforce weapons as "destabilizing." No recommendations are offered as to what targets the U.S. *should* threaten to strike in retaliation for Soviet first use, or what military capabilities would be required for these retaliatory attacks. A U.S. nuclear-war plan informed by the directives contained in the Pastoral Letter would have no direction or purpose. The credibility of an effective U.S. reprisal for the most serious Soviet aggressions would be severely undercut. This, in turn, could result in Soviet provocations that might precipitate nuclear war—the abhorrent act at the center of the letter's concern.

An appropriate no-first-use policy implies a willingness, if necessary, to resort to a justified second use. That justification, as noted, would be only to deter, balance, and correct for the possession or use of nuclear weapons by others. Under such circumstances, appropriate second use would be a justifiable moral action.

But no-first-use can also weaken deterrence if it results in a significant lessening of the possible risks confronting a potential aggressor. I would accept some such weakening, if necessary, but would prefer to see governments adopt this policy in a way that strengthened deterrence, i.e., by accompanying it with a credible ability to counter conventional attack with nonnuclear forces, and to alleviate the consequences of a nuclear conflict if deterrence fails. The Pastoral Letter condemns most possibilities for doing this on the basis of some very dubious strategic judgments.

Similarly, one wonders about the bishops' opposition to any efforts by the United States or NATO to gain nuclear superiority. If we knew clearly what superiority meant, this might be a reasonable position (although I would probably be opposed to it). But the fact is that superiority depends upon assumptions about the causes, conduct, and consequences of nuclear war—i.e., on the context in which one makes the estimate. If one has the unalterable conviction that any nuclear war would be the end of history, then the concept of superiority is meaningless. The only harm done by its proponents would be the misutilization of resources, or a very remote possibility that some people might believe in nuclear superiority to the extent they might be willing to risk or actually wage nuclear war; or, alternatively, that the potential enemy would believe in the superiority of the other side and be "provoked."

The letter of course does accept deterrence, but what it calls a "conditioned moral acceptance." According to the letter:

> . . . we cannot approve of every weapons system, strategic doctrine or policy initiative advanced in the name of strengthening deterrence. On the contrary, these criteria require continual public scrutiny of what our government proposes to do with the deterrent.
>
> Nuclear deterrence should be used as a step on the way toward progressive disarmament. Each proposed addition to our strategic system or change in strategic doctrine must be assessed precisely in light of whether it will render steps toward "progressive disarmament" more or less likely.
>
> Progress toward a world freed of dependence on nuclear deterrence must be carefully carried out. But it must not be delayed. There is an urgent moral and political responsibility to use the "peace of a sort" we have as a framework to move toward authentic peace through nuclear arms control, reductions, and disarmament.

The letter is correct in arguing that we should seek to reduce the nuclear threat in the time afforded by deterrence. (I have made this same point for many years.) However, to put pressure on the government to justify every "strategic system or change in strategic doctrine" in terms of its contribution to the nebulous concept of "progressive disarmament" and to achieve quick results in Geneva would place the United States at an enormous disadvantage in negotiating with the Soviets. In essence, U.S. negotiators would labor under a new timetable, while their opposite numbers would not. If the antinuclear groups in the United States and the other NATO countries pressure their governments for quick results in the INF and START talks, they make progress less likely by making the negotiations hostage to Soviet intransigence. When the talks flounder on genuine mutual disagreements between the two parties, the Soviets can nevertheless publicly accuse the U.S. and NATO of "lack of good faith." The letter would then seem to advise that the U.S. make some *unilateral* reduction in its forces in order to induce a reciprocal Soviet action. As noted above, this displays a gross ignorance of the history of the U.S.-Soviet arms competition over the last ten to fifteen years.

Furthermore, it is simply not at all clear that the aim of negotiation should be disarmament "for its own sake," as opposed to arms control, to make the world safer. It ought to be quite clear that to cut down to very small forces on both sides could be quite dangerous (perhaps making them more vulnerable than present arsenals and less capable of supporting severely damaging retaliatory attacks).

The Pastoral Letter is, however, clearly on moral and political *terra firma* when it notes that there must be no use of nuclear weapons solely or mainly against civilians. But as a strategist, I would add "except as a last resort or in very special circumstances." For example, imagine that (as in one of our *Gedanken* experiments) a powerful enemy nation destroyed a U.S. city just to teach us some kind of lesson, but otherwise used no additional nuclear weapons. We might not be willing to launch a major nuclear war in response, but might instead choose to retaliate according to the talionic law of an "eye for an eye and a tooth for a tooth," a doctrine that permits—even mandates—proportionate retaliation while it forbids escalation. A more likely situation would involve the withholding of U.S. forces during a war (probably submarine-launched missiles) for retaliation against Soviet cities if the Soviets attacked U.S. cities. But here again, they would be used only in a more or less talionic fashion—and only in the absence of other possibilities for punishing the individuals or group that had ordered the attack on U.S. cities. (In contrast, the bishops' letter even forbids "retaliatory use of weapons striking enemy cities after our own have already been struck.")

Living with nuclear weapons means making sure that they are never thought of as "just another" military option or "just another" way of resolving serious conflicts. But it also means that they are available as a credible last resort. This is what is generally called a war-fighting position.

As defined in chapter 3, war-fighting strategies, like deterrence-only strategies, advocate measures to reduce the possibility of accidental war; advocate the maintenance of survivable forces; and support the transmission of reliable "go-ahead" orders. But they add to this an emphasis on how the war is waged and how it might be terminated should deterrence fail. Thus, according to strategists' use of the term "war fighting," the Pastoral Letter's concern with avoiding cities, and with the moral question related to the use of weapons, makes it a war-fighting doctrine. This, of course, is not the same as accusing the bishops of wanting to fight a nuclear war (nor should this accusation be leveled against strategists who favor a war-fighting doctrine).

As noted, the bishops state that "we cannot approve of every weapons system, strategic doctrine, or policy initiative advanced in the name of strengthening deterrence." Neither can we. But we can approve of some of them and can give good reasons for supporting them. A categorical statement such as the above, without adequate amplification, reflects a sincere concern but an unenlightened simplicity—perhaps even an evasion of some central (however difficult) moral and strategic issues.

Similarly, it is counterproductive to issue a facile judgment that the United States must not have any weapons with counterforce ability because such weapons would be "destabilizing." The incorrect idea that the limited measures now being considered by the government to minimize damage are inherently destabilizing, dangerous, and falsely reassuring runs through the Pastoral Letter. In fact, given the bishops' injunction (and my belief) that civilians are not an appropriate routine target, enemy weapons *should* be targets and the system must be designed for the task. And if the lives of Soviet citizens cannot be endangered for the sake of deterrence, then surely the U.S. government has a related moral obligation to protect its own citizens from nuclear attack —through programs like the civil defense measures I advocated in chapter 8. (The clear moral imperative to provide adequate protection for civilian populations is one of the issues the Pastoral Letter delicately sidesteps.) "Damage limitation" can be sought through a combination of offensive nuclear forces (e.g., the MX ICBM, the Trident II submarine-launched ballistic missile) and strategic active and passive defenses (e.g., ballistic missile defense, civil defense). These are the very systems the bishops oppose (or at least fail to endorse), and in doing so they create a significant contradiction in their position. The destabilization effect, if any, might or might not be significant, but might still be morally and politically acceptable as an unavoidable cost of preparing for the possible failure of deterrence.

The letter seems to endorse Schell's view that a nuclear war would threaten "the existence of the planet." It is conceivable that it would, but we have no clear evidence to that effect. As far as we can tell, the created order has survived even greater stresses than nuclear war as a result of natural forces. It does not improve the safety of the world to terrify (via distortion of the dangers of nuclear war) one side (ours) into weakness or even submission.

Most of the problems I have with the document stem from seemingly uncritical acceptance of many currently popular (and seemingly plausible) but largely emotional arguments. Because of its tremendous influence and prestige, the Church must be prepared to reconsider technical and strategic judgments that are either unproven or clearly wrong, and to defend as strongly as possible those moral canons on which there can be no compromise. (The authors of the letter made a commendable effort to do this by writing that "the application of moral principles does not have, of course, the same force as the principles themselves and therefore allows for different opinions.") It would be immoral and unwise to jeopardize our national security interests and the values of most of the world on the basis of a strongly held but emotional evaluation of basically technical and strategic issues. However sincere, the accuracy of these strategic judgments is at best uncertain, at worst incorrect.

The Pastoral Letter seems to me to come too close to jeopardizing U.S. and global interests. If it is intended as an enduring statement on war and peace in a nuclear age, then the implications of war-fighting doctrines and enhanced strategic defenses must be considered much more carefully and objectively than the bishops have done so far.

I am not suggesting that the Church should take my judgments (or those of my colleagues) as authoritative. I am suggesting only that many of these issues are more controversial or uncertain than many of the assertions by the bishops would seem to indicate. My views are almost certainly not *wrong* on any of the issues raised here (I have carefully restricted my comments so that I can make this remark quite responsibly), though others might not agree that they are entirely right. I believe the bishops also must have a high level of integrity in trying to arrive at defendable positions and must be prepared to drop or qualify positions that are undefendable, however fashionable they may be (particularly in liberal and leftist circles). The message of the Church leaders must be valid and persuasive enough on its own merits to impress even those outside the circle of "friends and relatives" in the "peace" movements and other dovish or individuals and groups.

Assuming the intention of the bishops is not to preach to the converted, they must offer more persuasive arguments to back up the moral validity of their position, or they must be more guarded and limited in their strategic and political judgments. For example, the letter's argument that the costs of the "arms race" divert resources that could be better spent on curing the world's social and economic ills may be true, but the same could be said for money spent on tobacco, liquor, space exploration, or anything else. If we are to have a morally acceptable posture, it is most unlikely that it will come for free; it is more likely that it will cost more: intellectually, financially, and perhaps politically.

I do not challenge the right or even the duty of the Church to take positions that are controversial in secular terms but well founded on religious terms. Accordingly, if the Pastoral Letter supports the concept of "no first use" and "no targeting of civilians" (or, better, "no *routine* targeting of civilians") but recognizes the security value of deterrence, then I think that the Church will be endorsing policies that are strategically as well as morally supportable, even if it might mean a more expensive and complex defense establishment for the United States and its allies. A great opportunity to seize the moral high ground would be lost if, in the long run, the bishops' efforts came to be seen as just another (misguided) commentary on the strategy and tactics of nuclear war.

A Long-Range Antinuclear Policy

The position I have taken for many years on the moral aspects of nuclear use (and nonuse) comes as close as any to putting nuclear arsenals into a proper ethical context. The strategy and tactics of nuclear war become secondary (but still very important) concerns. The primary focus is on the rationale for maintaining a force of nuclear weapons: the only justification for ownership is to deter, balance, or correct for the possession or use of nuclear weapons by others.

"Deter, balance, or correct" is a very important phrase, and one I use often. The first word, of course, means to dissuade by terror, and in fact "deter" might be changed to "dissuade" because we are interested in dissuading by any means that work, that are appropriate, and that will do the least amount of evil. So in many cases we may choose methods of dissuasion that are not by terror (e.g., by confronting the enemy with the fearful prospects of military defeat, revolution at home, or the "cost ineffectiveness" of his own attacks).

There are all kinds of subtle ways to use nuclear weapons. For example, by advocating no first use the United States is also giving up the right to *threaten* to use them in certain circumstances. We would, of course, still need to have the capability to enforce the ban on no first use; for this, a credible capability to initiate thermonuclear war would be required—a concept many people would deplore. But whatever limited bargaining power in crisis diplomacy may be lost by renouncing first use is regained in the morality of our effort to limit the influence of nuclear weapons (their actual and potential use) to as small a sphere as possible. We seek to eliminate any fringe benefits of simple possession, especially by smaller countries. But we have to have them, to "balance" the nuclear power of others.

By "correct" I mean that if somebody actually uses nuclear weapons, then something has to be done. The most basic correction is retaliation designed to make clear that an aggressor cannot get away with a violation of the moral principal of no first use, then to limit and alleviate the damage (to ourselves as well as to our opponents' populace) through U.S. civil defense, ballistic missile defense, air defense, and very careful and limited (in many cases "limited" only in the sense of less than all-out) second use of nuclear weapons against the aggressor.

I am very strongly in favor of the United States issuing two basic statements: First, we will not be the first to use nuclear weapons against a nonnuclear power. Second, sometime between 1985 and 1990 we will adopt a full no-first-use policy—that is, we will not be the first to use nuclear weapons under any circumstances. (We might well include in this statement bacteriological and perhaps chemical weapons.) The United States should make clear that the reason for a five-year transition period, rather than an immediate policy change to a general no-first-use policy, is that it has certain obligations to its European allies, particularly to the West Germans, and that it does not intend to modify these obligations without consultation. In private, the U.S. should make clear to the West Germans and other Europeans that the announcement will be closer to 1985 than 1990. The reason for this rather tough policy is that the mere announcement is not likely to induce any great improvements in West European conventional forces to compensate for the withdrawal of the first-use guarantee; there would have to be large and explicit pressures as well. These pressures could, of course, lead West

Germany to "go nuclear." This is a result that most people, including the author, feel could be very risky. While I actually feel that in the long run it might not be a bad idea for West Germany to be nuclear-armed, the transition to that state is so fraught with perils, dangers, and political complexities that this is one of those cases where you just cannot go from here to there without exercising great restraint, and without very complete consultations regarding the intentions on both sides.

Alternatively, the pressures on the West Germans to go neutralist would be very strong. And this might, in fact, turn out to be their best choice (in terms of their national interest), particularly if they were politically able to combine this neutralism with a very large conventional force so that their neutrality was not likely to be violated without a major conventional confrontation or an escalation to Soviet use of nuclear weapons. At the same time, such a large conventional force would be safe from the viewpoint of the rest of Europe because almost the only countries the West Germans could attack would have nuclear weapons or nuclear support (by the Soviet Union, France, England, or even the U.S.). Despite any no-first-use pledges by these powers, their nuclear arsenals would appear quite dangerous to any West German leadership contemplating an attack or creating great political-military pressures against any European country.

The United States should make clear that its no-first-use policy is not a disarmament proposal. In many ways it is a proposal for additional armaments. I find one of the weakest parts of the widely discussed article by Robert McNamara, McGeorge Bundy, George Kennan, and Gerard Smith*—in which they advocate no first use—is their contention that the policy would diminish the need to modernize our nuclear forces, and that the cost savings could be used to improve our conventional posture. In other words, they suggest that the United States would need only a relatively inexpensive and pure second-strike nuclear capability. The article does note that we could not be sure that the Soviets would hold to their pledge not to be the first to use nuclear weapons (the Soviets have already made such a pledge, as have the Chinese) and we, of course, should realize that they could not be sure that we would hold to ours.

If the United States and the NATO alliance do adopt a declaratory second-use strategic position, then, in addition to strengthened conven-

---

* "Nuclear Weapons and the Atlantic Alliance," *Foreign Affairs*, Spring 1982, pp. 753–68. (The authors are sometimes referred to collectively as the "Gang of Four.")

tional forces, it would have to be combined with some concept of *lex talionis** for dealing with any Soviet violation of the no-first-use pledge. If it is not, then the conventional force buildup is essentially invalidated by Soviet nuclear superiority (or "parity-plus"). It would be much better to be able to match the Soviets in any escalation (conventional or nuclear) or to threaten them with a possibly disarming strike if they provoked us too bizarrely. Otherwise, why would a "second-use" guarantee be any more deterring to the Soviets than the present U.S. first-use threat that Bundy et al. do not consider to be credible? A no-first-use policy is not going to work well enough (that is, be sufficiently credible) without a convincing military program behind it. Our guarantee to NATO and especially to West Germany will not ring true, and our professed revulsion toward nuclear war will not ring true, unless we are genuinely capable of punishing violators of the no-first-use doctrine.

The authors of the *Foreign Affairs* article recognize that this policy would be received with the greatest apprehension by West Germany, and they argue that this apprehension can be eased by reassuring the West Germans that while renouncing nuclear retaliation to a conventional attack, we would still retain the policy with regard to a nuclear attack. However, if West Germany were actually faced with defeat in a conventional conflict, it is possible that the American leadership might change its mind—and would need more than a simple second-strike force in order to make or carry out a first-strike threat. By the same token, if the United States had a "deterrence-only" posture that relied exclusively on a massive and indiscriminate response to nuclear provocation (which would clearly entail the destruction of the United States in return), it is just not credible that the United States would retaliate with nuclear weapons, even in a response to the Soviet use of nuclear weapons on German soil.

Nonetheless, if I were a West German I might not feel terribly frightened, because like most West Germans, I do not believe that the Soviets have much desire or are willing to take any major risks to take over the country (and even less to reunite East and West Germany), but I certainly would not feel reassured. Indeed, the proposal formulated by the "Gang of Four" seems to suggest some unilateral U.S. disarmament combined with some withdrawal or weakening of the U.S. guarantee. This perception is likely to be widespread among West Germans and is therefore likely to *increase* existing fear in West Germany of a Soviet attack or Soviet pressure.

* See page 214.

There are very few people in West Germany, even outside of the "peace" groups, who believe that much of West Germany would survive even a limited tactical nuclear war; the most likely West German reaction to the use of nuclear weapons, or credible threat of such use, would be to surrender. In fact, I would argue that much of Western Europe has a policy that could best be characterized as "preemptive surrender" in the event that a Soviet attack seemed imminent during a crisis. To compensate for this, at least to some degree, I believe that it is urgent for Western Europe to improve the effectiveness of its conventional arms. Before adequate conventional armament would be achieved, however, there would have to be fairly tough U.S. pressures, and there would have to be a significant belief that there is still a U.S. nuclear guarantee to Western Europe that is "not incredible."

The political cohesion of the NATO alliance would be greatly improved if nuclear weapons—their kind, numbers, siting, contingency plans, and so on—were no long a major and chronic bone of contention. A declaration of no first use, coupled with credible nonnuclear defense options within Europe and improved nuclear capabilities, would alleviate much of the internal dissension that has existed on this issue over the past decade at least. In addition, anything that helps reduce the reliance on nuclear weapons helps reduce the potential for nuclear war.

For the past twenty years I have been concerned with how best to reduce this potential for nuclear war. Certainly the instincts of the utopian (Schell) and religious (bishops) antinuclear advocates are right; even the revised "Establishment" position (Bundy, et al.) comes much closer to the mark than their positions two decades ago. But it was around that time that I suggested a "long-range antinuclear policy" (with no first use at its core), which I still believe (indeed, am more convinced than ever) constitutes the only politically, morally, and militarily defensible nuclear policy for the long run. Specifically, a nuclear policy should accomplish the following objectives:

1. It should make nuclear weapons be and seem to be virtually unusable —either politically or physically.
2. In particular, it should prevent nuclear intimidation (except for the threats needed to preserve nuclear deterrence).
3. It should decrease the prestige associated with owning nuclear weapons (perhaps by limiting proliferation to regional military organizations whose purpose would be to provide for a nonnational tit-for-tat retaliation, or by a more or less explicit U.S. or Soviet talionic guarantee to various nonnuclear areas).
4. It should limit proliferation without necessarily freezing the nuclear

status quo (nations should not be put into an unnecessarily vulnerable security position).

5. If nuclear weapons are used, it should limit the damage that is done —it should not rely on deterrence working perfectly (an explicit policy of proportionate retaliation, or *lex talionis,* would impose intrawar limits on escalation, while strategic defenses would provide an important degree of direct protection against attack).

6. It should be competent (i.e., resilient and flexible enough) to withstand crises, small and even large conventional wars, and even some nuclear breaches and violations.

7. It should be responsive to national interests, sentiments, and doctrines, and should be negotiable.

8. It should improve current international standards, but should not require thoroughgoing reform (a responsible nuclear policy is not a moral mission to redeem mankind, but a program to reduce the risks and costs of war among nation-states).

9. It should be potentially permanent (i.e., not designed as a transitional arrangement) and yet be flexible enough to constitute a hedge against events and opportunities in both negotiation and operation—it should allow for major or basic developments and changes.

Some of these objectives may appear to be almost as utopian as those Schell-like visions I have rejected as unrealistic and unrealizable. But in fact, most of them should not present insurmountable obstacles, and if such arrangements are successful, they could limit the further spread of nuclear weapons and increase the credibility of a talionic response to a nuclear provocation—and therefore the deterrence of provocation. Alternatively, if deterrence failed and weapons were used, the result would not inevitably be Armageddon but would be limited to whatever destruction was entailed in the tit-for-tat exchange. After that, there would presumably be a return to some previous status quo.

The objective of proportionate retaliation is to bring the violence to a rapid conclusion and to create precedents that prevent recurrence; it is not to determine who was "right" or to consider other abstract points of law. (In this sense, the talionic rule is conceptually akin to the role of many UN peacekeeping operations.) As has been true for *lex talionis* arrangements from the most primitive cultures on, their primary objective is to restore equilibrium.

And yet there are certain objections that can be made to the basic tit-for-tat doctrine. For example, the idea of "an eye for an eye" might be manifestly unjust when it actually means "a city for a city," and when the city attacked in retaliation is inhabited by persons with no special responsibility for the initial nuclear attack. Serious ethical and political questions are raised, which depend in part upon empirical, analytical, and technical considerations, such as what response is proportionate or whether a talionic doctrine would be more unjust or unstable than alternative doctrines. The injustice and other defects of inflexible tit for tat must be compared with the possible infeasibility, risk, or even immorality of some counterforce or massive city attacks, and other more or less flexible, ambiguous, or unpredictable doctrines, as well as the possible consequences of not retaliating at all or not retaliating in a manner that deters further attack. None of these questions can be answered simply or dogmatically. In any case, some allowance might be made for responsible authorities to avoid at least the most rigid kind of city-for-city retaliation.

To prevent misinterpretation of an intended tit-for-tat response, slight underescalation might be advisable. Furthermore, within the limits of technical capabilities and political circumstances, it might be advisable to select on an ad hoc basis a different kind of target in retaliation—say, an isolated military installation for a city—or indeed, in certain special circumstances, to use nonnuclear means to enforce the lesson that nuclear weapons are not to be used.* In practice the *lex talionis* need not be absolutely inflexible in order to be effective; the doctrine of "justifiable reprisal," often invoked in the nineteenth century, offers better analogies than does the literal practice of eye-for-an-eye retaliation among some primitive peoples.

This long-term antinuclear policy would build upon the common revulsion at the thought of using nuclear weapons. For example, "imperialism" and "racism" were operative and even acceptable theories several decades ago, but they are now "out." I believe it is not impossible that sometime in the next few decades the illegitimate possession of nuclear weapons (i.e., possession other than to deter, balance, or correct for their use by others) will seem equally reprehensible. Antinuclear sentiment will flourish (but not as it does today in an emotional, partisan, or counterproductive manner). Creation of a durable antinuclear taboo requires the joining of the popular psychological and moral

---

* Even in the event cities were to be destroyed under the talionic rule, precautions could be taken to lessen the human suffering by a significant degree. For example, evacuation of the targeted cities could be permitted prior to their attack. (In some conceivable conflicts, enraged survivors in the aggressor nation might attempt to overthrow the regime that had caused them such tremendous hardship. The prospect of this kind of internal disorder could be a more powerful deterrent threat to the regime than that of counterpopulation retribution.)

abhorrence of nuclear weapons with the policies outlined above. People must be not only opposed to nuclear war, but ready to foreclose the possibilities for nuclear intimidation and to correct for nuclear use. Otherwise, the widespread fear of nuclear war will be manipulated by parties not sharing the taboo to forestall opposition to military aggression. Without tying antinuclear sentiment to a sound long-term antinuclear policy, there is a risk that a nation that legitimately had to resort to nuclear weapons for its self-defense might become a pariah.

No scheme can completely assuage the multifarious fears that arise as a consequence of the existence of nuclear weapons; for even in the case of total disarmament, fears will remain that the weapons might again come into existence or that disarmament was not complete. Nor will any scheme completely eliminate nuclear weapons from the calculations of statesmen, and in fact, it might be undesirable to try to go that far. The major, perhaps sole legitimate function of nuclear weapons should be to deter—to answer the threat or use of nuclear weapons. Aside from this the world should psychologically be *relatively* close to a situation in which nuclear weapons did not exist.

The reason for the emphasis on the word "relatively" may need clarification. A widespread belief that the world was completely nonnuclear would clearly result in some undesirable effects. For example, nations today are very careful in nonnuclear confrontations simply because they fear escalation to nuclear weapons (or the procuring of nuclear weapon systems by their opponents). These inhibiting fears have some desirable consequences. But we may be able to eat our cake and have it too. Such fears of nuclear escalation or production will still exist despite measures that successfully limit proliferation and reduce the likelihood of the use of nuclear weapons.

A declaration of no first use (as part of a long-term antinuclear policy) is not a panacea, nor is total trust in religious credos, nor is an underlying faith in the ultimate goodness and rationality of mankind. Spiritual and moral values certainly inform practical and secular ones, but they cannot substitute for them. Even the most devout and pious practitioners have to live in the real world, and almost all religions accept the basic notion that the Lord helps those who help themselves. In terms of seizing the moral, political and strategic high ground, this means that a nation, in addition to relying on God's good will and aid, is entitled to use force in defense of the lives, property, and values of its citizens if it is attacked. In fact, we would argue that anything less would be immoral and irresponsible behavior.

APPENDIX

A Framework for Central Nuclear War Issues: Basic Comments, Concepts, and Contexts

This appendix contains a series of five outlines I have used in numerous lectures given to experts and laymen alike. Together the outlines constitute a framework and checklist for representing and analyzing central nuclear war (CNW) issues. Their purpose is to provide a common language and context for discussing CNW that is limited enough to be manageable, simple enough to be clear, precise enough to be efficient, and comprehensive enough to cover most of the issues and topics presently of highest priority or interest. In this way, they might improve military planning for CNW and raise the level of general discussion about these problems.

The outlines also summarize many of the most important points made in this book. (In fact, chapters 6 and 7 are essentially elaborations of outlines 4 and 5.)

#1: How Is Central Nuclear War (CNW) Different? (i.e., What Has Changed Dramatically?)

I. **Potential for great destructiveness and terror**
   A. Potential rapidity, scope, and overwhelming character
   B. Special threatening, even schrecklich (terrifying), character of many weapon effects
   C. Likely inability of active and passive defenses to provide reliable high-level protection; thus distinctions among "deterrence," "denial," and "defense"
   D. Complexity and uncertainty of overall analyses
      1. Phased implementation of various U.S. and Soviet defense programs
      2. Wartime performance of U.S. and Soviet forces
      3. Immediate physical and psychological consequences of all relevant weapon effects (e.g., blast, nuclear and thermal radiation, electromagnetic pulse, dust, ground shock, debris)
      4. Possibility of tactics designed to exploit or diminish these effects for narrow military purposes, for coercion, or for intra-war deterrence
      5. Other effects of the size and "shape" of the war, including how the war is fought and terminated
      6. Postattack problems of short-run survival and reorganization, and eventually, long-term economic and social recuperation
      7. Impact of postwar economic, political, military, and social context, both domestic and international
      8. Various health and genetic problems over the short, medium, and long term
      9. Potentially catastrophic environmental aftereffects often associated with nuclear war (such as a new ice age or the destruction of the ozone layer)
      10. Surprises (which are more likely and more significant with larger attacks and less "controlled" wars)
   E. One result is an unprecedented emphasis on deterrence and the psychology of deterrence (and the issues of credibility associated with deterrence)

II. **Some common attitudes and strategies**
   A. Psychological or other rejection (e.g., "CNW is unthinkable")
      1. "The survival of mankind is at stake"
         a. CNW inevitable would be the end of history

    b. It is best to think that it would be

    c. Or at least might be

  2. "War never pays": if this aphorism wasn't true before, it certainly is now

  3. Deterrence ("dissuasion through terror"): e.g., the threat of "mutual assured destruction" (MAD) is the best bet

  4. The virtue of an appropriate* "no-first-use" concept: e.g., the position that possession and use of nuclear weapons should be taboo except for balancing, deterring, or correcting for the possession or use by others

B. But there can still be a need for *insurance;* some also want *reinforcement of deterrence by means of rational dissuasion* (by further terrifying the enemy or confronting him with the fearful prospects of military defeat, revolution at home, or the "cost-ineffectiveness" of his own attacks)

    This is a rejection of "deterrence only" and an emphasis on "war fighting." Which raises all the complexities and uncertainties confronting "nuclear-use theorists"

III. Possibility of radical differences in basic doctrine

A. Perceived need (either paranoid or real) for ideological (or at least holistic) policy making; incrementalism and compromise are regarded as basically unacceptable or infeasible

  1. "War fighting": great concern about deterrence but with some emphasis, if deterrence fails, on how the war is fought and comes out

  2. "Deterrence only": little or no concern about "details" other than survival against a first strike. Reliable transmission of the emergency war order, initial targeting of civilian populations, and prevention of inadvertent war

  3. Growing importance of "nuclear pacifism"

B. Less extreme and more ad hoc doctrines, involving more or less "muddling through"

  1. Use of special concepts: e.g., talionic ("tit-for-tat") nuclear reprisals (now sometimes called "assured proportionate reprisals")

  2. Compromising the uncompromisable, often by allowing internal inconsistency

  3. "Lucking out" in avoiding nuclear crises or nuclear war, by "1" or "2" above, or by depending on ambiguity, uncertainty, compromise, or lack of pressure

* Note that no-first-use proposals differ. For example, the author's proposal is quite different from that of Messrs. Bundy, Kennan, McNamara, and Smith (sometimes called the "Gang of Four").

IV. Importance of arms control

    A. CNW is not likely to be a "zero-sum" game (where the gain of one side is the loss of the other)

    B. Relatively successful role to date of self-restraint in restriciting access to nuclear weapons, their use, and even the role of "nuclear diplomacy"

    C. Need for implicit and explicit (or both "natural" and "coordinated/negotiated") agreements

    D. Possibility of a stabilizing change in the structure of nuclear politics in about two or three decades

V. Reasons for going to CNW

    A. None (i.e., inadvertent war)

    B. Action to avert disaster, i.e., the consequences of not fighting may appear more "dangerous" or "disastrous" than CNW

    C. Retaliation or punishment of others for using nuclear weapons
    Note that none of the above reasons include ambition, positive gain, or the exploitation of special fleeting opportunities, though these could reinforce *"B"* or *"C"* above

VI. A prewar lack of emphasis on appropriate strategies and tactics, yet a special need for such emphasis (since there would be little or no time in a war to learn or experiment)

VII. See also Outline 2, "Some Basic and Introductory Comments on Current Nuclear Doctrines, Strategies, and Tactics," for more illustrations of discontinuities and dramatic changes

#2: Some Basic and Introductory Comments on Current CNW Doctrines, Strategies, and Tactics *

I. Some useful (though sometimes pejorative) † descriptive terms

    A. *Pacifism:* In the age of "social limits to growth" and megaton weapons pacifism is more defendable and politically palatable than it was in previous periods, but it still has many of the tradi-

---

* These basic comments pertain to the "revolution in warfare" because the degree of concern over them has increased dramatically since World War II.

† We wish to emphasize that almost all of the positions are—or should be—uncomfortable, even to those who advocate them.

tional problems and, perhaps, fewer of the traditional virtues. It may also be an action policy, but not a declaratory policy, e.g., "preemptive (or preventive) surrender" policies

1. *Religious or philosophical:* Taking literally the Christian injunction to "turn the other cheek" and to "love your enemies"; or it can be simply the commitment to nonviolence found in many other religions and philosophies
2. *"Cost-effective":* Various forms of passive resistance designed to frustrate an aggressor, which are believed to be more effective or practical than the use of violent resistance
3. *"Nuclear":* A willingness to use conventional weapons and violence but not nuclear weapons, even to avoid defeat and sometimes not even in reprisal for the enemy's use of nuclear weapons

B. *"MAD"* ("mutual assured destruction"): This position is simple, understandable, unaggressive, relatively inexpensive, and relatively acceptable politically and emotionally (because it seems so practical and moral—and even convenient)

C. *"NUTS"* ("nuclear use theories" or "theorists"): a relatively rational, logical, and theoretically defendable position, but one which also has some (potentially overwhelming) public-relations and technical problems

D. *"Gnostic":* More or less esoteric strategies and tactics that may, on the whole, be quite relevant and not unreasonable.

Examples include multistable deterrence; the use of "not-incredible counterforce first strike" threats for extended deterrence; "tit-for-tat" nuclear exchanges for retaliation or punishment; other "limited strategic options"; "calculating" war; some flexible (nuclear) response policies; and other types of limited war

E. *Loutish* ("lucking out"): Many tend to rely on ambiguity, uncertainty, or psychological avoidance, and perhaps compromise or appeasement—or just the lack of a threat—to avoid CNW. If carried too far, loutish policies can appear to critics to be stupid or brutish, or at least irresponsible. Nevertheless, dependence on lucking out (or better, "reinforced lucking out") may be hardheaded and realistic in terms of feasible current options and alternatives

Current examples include threats of nuclear escalation in the event NATO's conventional defenses fail; emphasis on "prevailing" in a "protracted" (months or years) CNW after massive countercity attacks; waging three-front (full-scale) wars with the Soviet Union without a lengthy period of mobilization; and threatening a nuclear defense of the Persian Gulf or a "horizontal escalation"

(i.e. precipitating a new crisis or conflict outside the immediate theater of war)

II. Less or more gnostic and/or loutish beliefs (i.e., "A" is less gnostic and/or loutish than "D")
  A. Some dependence on mobilization for deterrence and defense
  B. Great dependence on Soviet caution and prudence
  C. No clearly visible or intolerable threat
     Examples:
     1. The views of the West German peace movement
     2. Current Japanese "pacifism"
  D. "Preserving security is not our responsibility" (an attitude based on wishful thinking, illusioned analysis, or a feeling of helpless fatalism)
     Examples:
     1. Pre-WW II Czechoslovakia and Poland
     2. U.S. dependence in the nineteenth century on the protection provided by the British navy
     3. Most of the world today

III. Need to distinguish many degrees of openness, explication, likelihood, or "officialness"
  A. In declaratory policies
     Example:
     1. NATO public and official policy: use nuclear weapons to avert defeat
     2. The open secret: preemptive or preventive accommodation or compromise by European countries in the event of Soviet attack or threat of attack
     3. Officially secret policies: most NATO plans
     4. Genuinely secret or covert policies: some unilateral plans of NATO members
  B. Action policies
     1. Officially planned
     2. Various estimates
     3. Likely responses
     4. Possible responses

IV. The difficult role of the just war doctrine in the context of nuclear war
  A. Traditional content
     1. Presumption against the use of force
     2. Competent authority

      3. Just cause
      4. Right intention
      5. Last resort
      6. Probability of success
      7. Just means
         a. Discrimination
         b. Proportionality
  B. With the possible exception of criteria "6" and "7," the just war doctrine can be used to justify current war-fighting policies and strategies (although the Catholic bishops have issued a pastoral letter challenging this interpretation)
  C. As a result, much new Church thinking appears to critics to be morally less tough, informed, and practical than it has been traditionally. Advocates of the pastoral letter disagree

V. Importance and relevance of *Gedanken* (thought) experiments
  A. Hypothetical experiments that are not intended to be carried out, but could conceivably be done
  B. Can be usefully applied to CNW issues
      1. Forces one to think through CNW issues and consider the consequences of responses to different situations (tries to capture drama and tension of the moment)
      2. Permits consideration of alternative scenarios and policy options
      3. A useful (though far from perfect—indeed clearly inadequate) substitute for experience

VI. Need for apparent double negatives for precision: e.g., "not incredible" ("not incredible" is an example of what rhetoricians call a "periphrasis")

#3: How Is CNW the Same?—or How Much Has Not Changed So Greatly?

I. CNW can still be thought of as an experience and not the end of history. One must worry about "Pearl Harbors," "Munichs," and the possible outcome of a war, as well as Armageddons
  A. There can still be important differences between victory and defeat, and while victory can be costly, it need not be Pyrrhic (and Pyrrhic victories are not new, either)

Appendix

B. Some credibility of both Type I and Type II Deterrence (e.g., multistable deterrence) is necessary

II. There are still rational and moral reasons for threatening (nuclear) war—or even waging it—and, of course, irrational, immoral, "a-rational," and amoral reasons as well. That is, deterrence may fail either deliberately or inadvertently
   A. The possibility of intense confrontations and escalation from peace or limited war remains. This would be true even if "I" above were not the case—even more so if it is
   B. (Nuclear) war as an instrument of foreign policy or as "a continuation of policy by other means" is not totally obsolete, though it *appears* to be both a blunt and awkward tool and relatively obsolete

III. Possibility of different kinds of (nuclear) wars
   A. Eight basic prewar contexts
      1. Normal situation
         a. Soviet strike out of the blue
         b. Accidental war
      2. Tense crisis
         a. Soviet Union solves a problem
         b. A "less accidental" accident
         c. U.S. solves a problem
      3. Basic change in the strategic context
         a. Deterioration of international relations
         b. Arms-control developments
         c. Technological breakthrough
   B. CNW may not be either total or an uncontrolled "wargasm"
      1. Limited nuclear war and flexible response are possible, at least in the initial stages of a conflict
      2. Even high-level nuclear war may last days or weeks (or possibly longer)
      3. Calculations, tactics, strategy, military reserves, and postwar forces may all affect the course of events

IV. Continuing or increased importance of:
   A. Strategy and tactics
   B. Bargaining, communication, and deterrence: preattack, intrawar, and in war termination
   C. Various thresholds and firebreaks (the "escalation ladder")
   D. Preattack preparations and various improvisations
   E. Active and passive defenses

    F. Intangibles (skill and quality, determination, courage, commitment, "rationality of irrationality," and willingness to "play chicken")

    G. Recuperation and mobilization capabilities

    H. Long shots and surprises

V. Possibility (and importance) of formal declarations of war and (more or less) "phony" wars

VI. Technological, economic, cultural, and political developments can still change the strategic environment

    A. Technological changes:

        1. Use of "smart" conventional and/or very low yield or special-purpose nuclear weapons (even if for only a limited number of targets) soon could make "surgical strikes" and limited central wars much more feasible

        2. Possible role of space-borne defense systems, and/or

        3. Various types of directed-energy weapons (e.g., lasers, particle beams)

    B. Economic change: Emergence of a truly multipolar world

        1. By 2000, there could be a trillionaires' club with eight members. The U.S. could have a GNP (in 1982 dollars) of about 5 trillion, Japan 3–4, the Soviet Union 2–3, China 1–2, and Brazil, West Germany, France, and Mexico between 1 and 1.5

        2. Such a multipolar world might be more stable than today's international system

            a. The victor of a two-power war cannot hope to attain world domination easily—or even maintain his prewar status

            b. Balance-of-power politics (as in Europe from 1815 to 1914) becomes possible and likely

            c. Such strategies as "MAD," "launch on warning," or the deployment of potentially uncontrollable forces become much less desirable and defensible

    C. Cultural changes: emergence of overlapping "Green," "social limits to growth," antinuclear, anti-American, and anti-industrial ideologies and movements

    D. Political changes: modification in, or change of, present alliances or governments

VII. A belief in simplistic solutions to the problems of war is likely to persist. Many now find these solutions more attractive than ever, particularly if they support a position characterized by self-righteousness, moral indignation, and an almost complete ignorance of

—and indifference to—"nuances" and "details." This is not new, though the tendency toward simplicity may have increased

VIII. In short, there is no automatic and mutual "overkill" or reliable stability (even stability demands eternal vigilance). Equally important (even if overlooked), the comparative quality and quantity of strategic power (even if much less influential and usable than it used to be) can affect diplomacy, the three kinds of deterrence, and crisis management, as well as war fighting. Finally, the enemy can still be more dangerous than the "arms race"

#4: Five Canonical Scenarios for the Outbreak of CNW *

I. **Surprise attack (more or less out of the blue)**
   A. Deliberate attack: constrained, unmodified enhanced, or environmental attacks (also note "IV.C" below)
   B. Inadvertent war: because of accident, erroneous information or judgments, unauthorized acts resulting from irresponsible or faulty behavior, "reciprocal fear of surprise attack," or other disaster-prone crisis dynamics.

II. **Early eruption to CNW from an intense crisis**
   A. Four typical "not implausible" outbreak crises (listed in order of their relative plausibility)
      1. East European crisis: e.g., an East German or Polish uprising, or a crisis in Yugoslavia
      2. Persian Gulf disaster: U.S. "compound" (horizontal) escalation or use of nuclear weapons
      3. Sino-Soviet war (or crisis), with U.S. intervention, and eventual escalation
      4. East Asian crisis involving escalation from a Chinese–South Korean, Chinese (People's Republic)–Chinese (Taiwan), North Korean–South Korean, or some "Vietnam"-type confrontation
   B. Soviet nuclear strike (preemptive or "therapeutic"), or more likely
   C. Soviet urban evacuation (there is almost no chance, under current conditions, of U.S. preemption)

---

* These five classes of outbreak scenarios provide a reasonable basis for planning U.S. central war and mobilization capabilities. They can also be used for educational and polemical purposes, and somewhat less effectively as "triggering scenarios" (i.e., a sequence of events that precipitates a U.S. or allied strategic mobilization).

1. Very likely this move touches off a U.S. evacuation (voluntary, prudential, or strategic)
2. There now is an increased chance of inadvertent war (see "I.B" above)
3. This may be followed by a Soviet or U.S. ultimatum and/or attack (there is a high probability that the attack will be constrained)

III. "Classic" U.S. Type II Deterrence
A. An attack on a U.S. vital interest (e.g., a Soviet attack on Western Europe)
B. Evacuation by the U.S.
C. Eventually a U.S. ultimatum and/or a constrained counterforce attack accompanied by an offer or an ultimatum to the Soviets

IV. A "protracted crisis" that does not get settled but eventually escalates to CNW because of:
A. Inadvertent CNW (see "I.B."), which may now be relatively probable
B. A scenario similar to II or III above
C. An erosion of the capabilities of military forces on alert, and/or ("treacherous"?) relaxation of tension, followed by an enemy attack
D. A protracted limited war, including SASUS (Space, Air, Sea, and UnderSea) wars, followed by an escalation to CNW
E. A sequence of: provocation—initiation of mobilization—mobilization war (or a declaration of war, followed by some limited violence and then by a more or less "phony" war or other "lull")—escalation to high-level CNW

V. "Mobilization war"
A. Following a crisis or other triggering event, a U.S. mobilization is touched off
B. The mobilization escalates to CNW in:
1. One year: the U.S. has, relatively speaking, a war-fighting posture (e.g., greatly improved command and control, massive civil defense, and increased readiness and usability of strategic forces)
2. Two years: a greatly improved war-fighting posture (e.g., a huge expansion of existing forces and procurement of more or less on-the-shelf weapon systems)
3. Three years: a relatively elaborate war-fighting posture (e.g., more rapid deployment of Trident submarines and initial operating capabilities for "new systems")

    4. Four years or more: ???

  C. *Or,* the mobilization war continues and does not escalate to CNW (but there may be a limited war with conflicting requirements for forces). There is eventually a dramatic change in the strategic balance as well as an acceleration of the "arms race"

It should be noted that a number of possibilities have been left out, including:

1. "catalytic" war (where a third country tries to touch off a war between the major powers);
2. preventive war;
3. war caused by personal ambition, pursuit of world revolution or world domination (or other positive national gains), exploitation of a special opportunity, excessive (or accurate) belief in war plans, or a deliberate attempt to improve the domestic political climate or distract popular discontent (although many of these motives could contribute in a secondary way to the decision to risk or wage war);
4. war resulting from the replacement of the present clearly cautious and prudent Soviet leadership by an "Alexander the Great," a "Napoleon," or a "Hitler";
5. war following a World War I–type scenario (i.e., an allied power gets involved in a crisis or war, possibly on its own initiative, which it hopes will be limited, but actually escalates through [in part] a most improbable or implausible chain of events).

Needless to say, the implausible events that started World War I did happen. And so might the other factors lead to central nuclear war. (History can be more cunning and perverse than any of the analyst's scenarios.) But scenarios based on these factors are less likely and/or important than the five canonical scenarios.

#5: Important Concepts and Issues for U.S. Strategic Mobilization

I. The three kinds of mobilization

  A. "Crisis": crash, improvised civil defense measures, and actions to increase the immediate combat readiness of strategic forces

  B. "Tactical": activities to increase the amount of manpower and materiel for an ongoing (or expected) noncentral war

  C. "Strategic": relatively huge increases in defense budgets, designed to shift dramatically the balance of military power (conventional and nuclear), both to deter war and to improve a nation's war-fighting capabilities. "Strategic" mobilization is more long-term than either "crisis" or "tactical" mobilization, and is used in "mobilization war."

II. The reputation (and probably the reality) of U.S. economic/technological power is still awesome to the Soviets. Since they seem likely to have considerable problems in the next decade, the threat of even a low level of mobilization war could be a deterrent to Soviet provocation. (Indeed, even the present proposed level of U.S. rearmament, while much lower than what we normally think of as a "mobilization war," is putting a lot of pressure on the Soviets.)

III. A properly designed U.S. strategic mobilization can be effective, probably much more so than most defense planners and analysts believe. Options for effective programs include:

A. Changing manufacturing techniques: tank turrets*

B. Finding additional sources: foundries for aircraft parts

C. Changing materials or other component specifications: piers

D. Changing design: MIG-25

E. Procuring from nondedicated sources: guns, tanks, airplanes, and ships in World War II; fallout and blast shelters from new and old inexperienced firms (but with construction experience); foreign procurement (from the NATO allies, Taiwan, South Korea, Japan, etc.), facilitated by the judicious encouragement of even small joint production contracts

F. Changing military tactics (including forces): use of mines instead of tanks for frontier or other defense

G. Changing military strategy: greatly increased emphasis on active and passive defense

H. Exploiting enemy and/or other uncertainty and ignorance: use of decoys and "Potemkin villages," deployment of systems of forces with incomplete or uncertain capabilities (e.g., early deployment of new and untested equipment or systems), and use of other extemporizations, improvisations, and duplicities

I. Using intuition, educated guesses, incomplete studies, "learn as you go" projects, and parallel experimental programs to compensate for lack of information and experience

J. Funding premobilization research and development, especially development of concepts and prototypes

K. Mobilizing the civilian economy effectively: use of preplanned "stop" and "trigger" orders and policies

*The examples cited for each option are described in chapter 7.

Appendix

L. Exploiting the many flexibilities of the current U.S. and other economies: CAD/CAM capabilities, vertical breakdown of the production process in order to exploit domestic and foreign comparative advantage, and other use of foreign capabilities and skills

IV. **Mobilization can be used to de-escalate a crisis (or at least to slow it down)**

V. **Precipitating crisis (or initiation of a mobilization) could also precipitate economic, financial, or political crisis; this, in turn, could greatly affect the course of events**

VI. **Extended alert, protracted crisis, limited war, or some other special situation could create both expected and unexpected problems and opportunities, many of which could interact significantly with any attempt or threat to mobilize**

VII. **An emphasis on mobilization capabilities—or an excessive fear of relying on or actually using such capabilities—can increase the likelihood of illusioned, wishful, or fearful planning or other mistakes**
   A. Excessive neglect of forces-in-being (siop and/or war-fighting capabilities)
   B. Excessive rigidity, appeasement, or other unwise crisis behavior
   C. Misestimation (either over or under) of Soviet willingness or technical capability to counter a U.S. mobilization
   D. Other misestimation (again either over or under) of domestic or foreign technological, bureaucratic, economic, or political problems or capabilities in mobilizing
   E. Critical unanticipated problems or opportunities (or predicted problems or opportunities that turn out to be quite different)

VIII. **Operational problems of a mobilization**
   A. U.S. military menu and priorities (e.g., force composition and procurement, with an emphasis on timing issues)
   B. Allied *vs* U.S. civilian and military priorities
   C. Internal military and civilian administrative and management issues (including civil/military relations)
   D. U.S. politics and "other internal inefficiencies": bureaucratic, armed service, cabinet, congressional, allied, and domestic

IX. **The possibility of mobilization should affect premobilization military policies**

A. System design and evaluation—no more "one-hoss shays"*
B. Research, development, test, and evaluation
C. Procurement policies (including the possibility of foreign procurement)
D. Alliance policies

X. Relationship of an adequate mobilization base to arms control
  A. The two are often psychologically or theoretically dissonant, but in practical terms it is important that they be reconciled
  B. Mobilization and arms control can also reinforce each other

---

*The example of a system that deteriorated totally when age and wear finally weakened it, each piece being equally as strong as the other. Military systems should instead be built to allow for retrofit of components, growth, and improvisation. (The term "one-hoss [horse] shay" comes from Oliver Wendell Holmes' poem, "The Deacon's Masterpiece.")

INDEX

Absolute Weapon, The (Brodie), 37
ACD (Arms Control Through
  Defense), 51
Acheson, Dean, 106
Afghanistan, Soviet invasion of,
  123, 161–62, 192n–93n, 198
air defense systems, 28n
alert status, 145, 158, 173
  as response to evacuation, 185
Alexander the Great, 146–47
Andropov, Yuri, 150
antinuclear movements, 47, 194–
  195, 197, 210
  see also nuclear freeze
Argentina, Falkland Islands
  invaded by, 130
arms control, 46, 51, 88–89, 190–
  204
  arms race and, 193–94, 196
  asymmetrical approaches to, 195–
  196
  breakout from, 27n, 162–63
  dangers lessened by, 25–26
  defense spending and, 199
  disarmament vs., 25–26, 198

favorable climate for, 37, 88
freeze movement and, 209
implicit codes of behavior as,
  192–93
international politics and, 191
long-term perspective needed
  for, 46–47
MAD as, 194–95
mobilization and, 162–63
multipolar world and, 200–204
multistable deterrence and, 119,
  195
national security and, 199
policy suggested for, 221–22
strategy vs., 194–95, 197
as substitute for direct military
  competition, 192
violations of, 162–63, 198–99
world government and, 24–25
arms race, 23–24, 26–27, 34–35
  arms control and, 193–94, 196
  civil defense and, 186–87
  multistable deterrence and, 118
  in Pastoral Letter, 217
Aron, Raymond, 120